ON THE NERVOUS EDGE
OF AN IMPOSSIBLE PARADISE

New Directions in Anthropology
GENERAL EDITOR:
Jacqueline Waldren, Research Associate at the Institute of Social and Cultural Anthropology, Oxford University, and Director, Deia Archaeological Museum and Research Centre, Mallorca

Migration, modernization, technology, tourism, and global communication have had dynamic effects on group identities, social values, and conceptions of space, place, and politics. This series features new and innovative ethnographic studies concerned with these processes of change.

Recent volumes:

Volume 45
On the Nervous Edge of an Impossible Paradise: Affect, Tourism, Belize
Kenneth Little

Volume 44
Nourishing the Nation: Food as National Identity in Catalonia
Venetia Johannes

Volume 43
Burgundy: A Global Anthropology of Place and Taste
Marion Demossier

Volume 42
A Goddess in Motion: Visual Creativity in the Cult of María Lionza
Roger Canals

Volume 41
Living Before Dying: Imagining and Remembering Home
Janette Davies

Volume 40
Footprints in Paradise: Ecotourism, Local Knowledge, and Nature Therapies in Okinawa
Andrea E. Murray

Volume 39
Honour and Violence: Gender, Power and Law in Southern Pakistan
Nafisa Shah

Volume 38
Tourism and Informal Encounters in Cuba
Valerio Simoni

Volume 37
The Franco-Mauritian Elite: Power and Anxiety in the Face of Change
Tijo Salverda

Volume 36
Americans in Tuscany: Charity, Compassion, and Belonging
Catherine Trundle

For a full volume listing, please see the series page on our website:
https://www.berghahnbooks.com/series/new-directions-in-anthropology

On the Nervous Edge
of an Impossible Paradise

Affect, Tourism, Belize

Kenneth Little

berghahn
NEW YORK · OXFORD
www.berghahnbooks.com

First published in 2020 by
Berghahn Books
www.berghahnbooks.com

© 2020, 2024 Kenneth Little
First paperback edition published in 2024

All rights reserved. Except for the quotation of short passages
for the purposes of criticism and review, no part of this book
may be reproduced in any form or by any means, electronic or
mechanical, including photocopying, recording, or any information
storage and retrieval system now known or to be invented,
without written permission of the publisher.

Library of Congress Cataloging-in-Publication Data

A C.I.P. cataloging record is available from the Library of Congress
Library of Congress Cataloging in Publication Control Number: 2019042651

British Library Cataloguing in Publication Data

A catalogue record for this book is available from the British Library

ISBN 978-1-78920-646-3 hardback
ISBN 978-1-80539-344-3 paperback
ISBN 978-1-80539-452-5 epub
ISBN 978-1-78920-647-0 web pdf

https://doi.org/10.3167/9781789206463

For Harry
(1947–2014)

But it wasn't magical [realism] to me ... if you live in the Caribbean you know that reality is wilder than the craziest fiction.
—Marlon James, *New Yorker Radio Hour*

The task approached in tenderness and faith is to hold up unquestioningly, without choice and without fear, the rescued fragment.
—Joseph Conrad, *The Children of the Sea*

Contents

List of Illustrations	viii
Acknowledgments	ix
Introduction. Writing Stories of *Make-Belize*	1
Chapter 1. "For the Time is at Hand": Beast-Time *Somet'ings*	25
Chapter 2. Impossible Tropics	43
Chapter 3. Richie's Tourists	57
Chapter 4. Nowhere Paradise	75
Chapter 5. Belize Ephemera	86
Chapter 6. Belize Blues	105
Chapter 7. Parca's Picks	127
Epilogue. Belize Fabulations	162
Glossary	174
References	177
Index	187

Illustrations

Figure 2.1. *Amandala* newspaper article announcing the Raelians.   53

Figure 5.1. Belikin Beer coaster for Lighthouse Beer.   87

Figure 7.1. Parca's numbers and a Boledo ticket.   130

Figure 7.2. One of Parca's number books for 2013.   134

Figure 7.3. A page of numbers from Parca's numbers book, 2013.   135

Figure 7.4. One of Parca's number swirls, 2013.   136

Figure 7.5. The "Grinning Doukie Skeleton," Norris Hall 1974.   145

Figure 7.6. Gilroy "Press" Cadogan's Boledo picks for the week.   155

Acknowledgments

This book is the product of a lot of travel and many years of "deep hanging" in Belize during which time I have amassed a lifetime of debt. I am forever grateful to a great many people for their generous support, encouragement, and friendship, and I know now how much books really are shared accomplishments. This list is long. It has been my privilege to share ideas and to receive guidance and close readings of some of this work from a remarkable group of colleagues at York University. My thanks go to the members and organizers of the Department of Anthropology Working Paper Series for giving me the opportunity to present chapters during two workshops. A debt of gratitude goes to Shubhra Gururani and Natasha Myers for organizing these very productive and lively events. I thank my department colleagues who took the time to read and discuss chapters one and seven with me, including Naomi Adelson, Othon Alexandrakis, Wenona Giles, Carlota McAllister, David Murray, Maggie MacDonald, Albert Schrauwers, Daphne Winland, and Dan Yon. I must also single out Penny Van Esterik for her mighty advice on several editing issues and to thank her for her overall encouragement. I have also been privileged to learn from an inspiring number of past and present graduate students over the time of this fieldwork and writing, including Heather Barnick, Jenny Burman, Shelley Butler, Lisa Cooke, Meredith Evans, Marc Lafleur, Kris Maksymowicz, Siobhan McCollum, Karen McGarry, Mary Lee Mulholland, Elysee Nouvet, Maria-Belen Ordonez, Natasha Pravez, Effrosyni Rantou, Nadine Ryan, Janita VanDyk, Andrea Walsh, and Jamie Yard. Ines Taccone, Kaila Simoneau, and Evadne Kelly graciously acted as my discussants during one or the other Working Paper workshops. I am grateful to the three of them for taking precious time out of their busy studies to make such wise and useful comments on my work. The author and publisher would also like to thank the Faculty of Liberal Arts & Professional

Studies, York University, Toronto, Canada, for the financial support it provided to this work.

It has been my privilege to be associated with the Centre for Imaginative Ethnography and to have had such continuous support of three of its leaders: Dara Culhane, and my York colleagues Denielle Elliott, and Magda Kazubowski-Huston. They have encouraged me to participate in several of their workshops and online collaborations. They also have very generously provided me with a radically open intellectual space in which to experiment with writing as well as opportunities to share my Belize stories with audiences broader than those I might normally have imagined. I offer special thanks to Lindsay Bell for acting as a discussant during one of these sessions. I also want to thank Stuart McLean for his remarks on a very early version of chapter seven. He shared them during an American Anthropological Association workshop organized by the Centre for Imaginative Ethnography in 2014. Many thanks to Kathy Bischoping for her useful notes on "Belize Blues."

Similarly, it has been my honor to be associated with the, now closed, Centre for Tourism and Cultural Change (CTCC), first when it was housed at Sheffield Hallam University and later when it moved to Leeds-Metropolitan University. I am heavily indebted to Mike Robinson, the Centre's Director. The CTCC may still be considered the most important intellectual hub for the study of tourism, ever, and that is mostly due to the care and effort that Mike Robinson and the faculty and staff put into it. The CTCC also created an exciting space for the dissemination of tourism studies by training a cohort of graduate students who have gone on to impressive careers of their own, including Valerio Simoni, who has become a close intellectual ally and friend. He has my many thanks for his encouragement and support. The CTCC organized a distinguished, if not exhausting, series of imaginative thematic discussions dealing broadly with themes in tourism such as movement, language, pilgrimage, ritual, spectacle, performance, affect, and materiality, to name a few. I was honored to be invited to many of these CTCC events. Several chapters of this book started as CTCC papers. The Centre always attracted a global range of researchers, practitioners, and creative arts professionals dealing broadly with travel and tourism studies, and it did more to encourage the critical study of tourism than any other organization that I can think of. I have made life-long colleagues and friends through the CTCC. It was at these events that I met Sally Ness, Alison Phipps, and Hazel Tucker, who have always remained interested and supportive of the work I have been trying to do. I am honored to call them cherished colleagues and friends and have learned much from them and their examples as critically innovative ethnographers of tourism. To them I owe a tremendous debt of gratitude. I owe a debt of gratitude to David Picard for his insights, friendship, and support while at CTCC events and beyond. In 2008, I was appointed a Visiting Research Professor at the Centre for Tourism and Cultural Change at Leeds-Metropolitan University, an appointment I held for five very productive years. I wish to thank the CTCC for that support as well. There is no replacing the CTCC.

The Critical Tourism Studies Network is another organization that shares a vision of how it may be possible to create social change in and through tourism practices, research, and education. I am particularly indebted to Kellee Caton for her support and for the hard work she does to organize CTS's biannual meetings. They attract a wide range of activists, artists, professionals, and scholars in allied fields of study that focus on tourism.

Some parts of this book were presented at several other workshops, including several meetings of the Canadian Anthropological Society, the American Anthropological Association, the Tourism-Contact-Culture Research Network, the Association of Social Anthropologists of the United Kingdom and Commonwealth, the European Association of Social Anthropologist, and the Royal Geographical Society. There have also been several tourism-related events that were instrumental in helping me develop my thinking on affect and tourism, including one in 2011 organized by The Tourism Working Group at the University of California, Berkeley, and the 2015 Worlds of Desire: The Eroticization of Tourist Sites event. There are too many workshop and session organizers, participants, and inspired interlocutors to thank here, but I will make special note of my long-standing collaborations with Susan Frohlick. We organized two very productive tourism panels through the Canadian Anthropological Association. Sue was supportive also by inviting me to participate in one of her Latin America Studies Association panels, co-organized with Valerio Simoni, in 2018. Some of that discussion and commentary has made it directly into this book. I am forever grateful for our collaborations and ongoing conversation and would like to thank Sue very much for her friendship and support. Julia Harrison has encouraged me from the very beginning to develop the kind of anthropology of tourism that inspires this book. I thank her for offering me such strong intellectual stimulation, encouragement, and support, as I have labored to rethink ordinary life in the new and extraordinary tourist state of Belize from a materialist, nonrepresentational perspective. Many thanks to Nelson Graburn for his example as the indefatigable scholar of tourism, discussion organizer, and for his support. I am also grateful to Noel Salazar for his interest in my work.

I was lucky to have had the services and use of the School of Theology Library at The University of the South in Sewanee, Tennessee. For four summers, and counting, Romulus Stefanut, Assistant Professor of Theological Bibliography and Director of the School of Theology Library, has made it possible for me to use their facilities while writing several drafts of my book chapters. I thank him and his staff for their interest and support. My time in Sewanee was coordinated to take advantage of the Sewanee Writers' Conference, which in itself has always been an inspiration.

I am indebted to Teresa Holmes, my wife and Belize research partner, for initially suggesting Belize tourism as a new research interest. Over the years of research in Belize, Teresa provided the historical frame for tourism there, fashioned out of her countless hours of hard work in the Belize Archives in Belmopan. But more than that, we spent many hours together daily listening to Love FM's *The Morning Show* while discussing local-level encounters, politics, and events so that we could set plans together that would initiate the daily rounds of fieldwork. Her patience, intellect, and encouragement throughout the research and writing years means the world to me, leaving me with a debt that I can never fully repay. My sincerest thanks also go to Elizabeth Graham. Liz is an archaeologist who has dedicated herself to years of ongoing field research in Belize. She is the friend who, when she was our colleague at York, first suggested Belize as the place to go if Teresa and I were truly interested in tourism studies. She saw what tourism was beginning to do to Belize, for better and for worse. After our initial family visit to Belize, we knew she was right. I am tremendously grateful for her initial suggestion and for her friendship and support throughout.

Of course, my deepest debt of gratitude is to the people of Belize. Special thanks go to Dawn Anderson and Pearl Cabral for helping me understand local coastal Belizean life in a way that otherwise would not have been possible. To Victor Ferrera, master local sailor and raconteur, a very special thanks for his friendship, sailing trips, and stories. Special thanks also go to Tom McTaggart for his long and sustained friendship and wonderful Belizean hospitality, and to Fred Mueller, indomitable logician, gamesman, and local phrase turner, for making me explain myself in a way that local Belizeans would understand. To Clay Robarchek and Carole Robarchek, expat cultural anthropologists in Belize, I thank for their personal kindness, intellectual example, spirit of adventure, and overall support. To Steve Shaw goes a very special thank you for his longstanding friendship and support and for providing me for many years with a roof to put over my head while in Belize. To Norman Leslie, respected village elder and successful local tourism entrepreneur, go my deepest thanks for his hospitality, honesty, and support over the years. My respect for Norman is deep. And to the members of the "Placencia Coffee Club," my sincerest thanks for early morning stories, gossip, hard talk, and general hilarity and camaraderie that daily buoyed my spirits. This is also to acknowledge that I still owe the group three bags of "Gallon Jug," along with a big debt of gratitude. I would also like to thank Brice Dial, Greg Duke, Rita Duke, Harry Eiley†, Bob Flindell, Sharyn Flindell, Lita Krohn, Devon Leslie, Margaret Martinez, Jake Roberts, Ellis Rojo, Mary Toy, Marsha Trent, Hector Tut, Froyla Tzalam, Dave Vernon, George Westby, Denise Williams, and Wendy Williams. Each in their own way, whether they knew it or not, has helped me over the years to understand Belize, tourism, and storytelling better, for that I am truly grateful. Marie, Ines, and Ruby, I can't thank you enough for your continuous good cheer, hospitality, local knowledge, and friendship.

I would also like to extend my sincerest thanks and deepest appreciation to the five scholars who reviewed the prospectus for the book and who were so encouraging and excited about the project. Your comments sustained me. To the two anonymous reviewers, whose careful, smart, and supportive reading of the book has helped me to improve it, you have my deepest thanks for your commitment to this project and for the good will and care with which you attended to the writing.

Finally, to Teresa and my son William, this book would not have been written without your support and love. You were both the most enthusiastic initial supporters of the project and you have kept me grounded and focused throughout. Your presence made a huge difference in the work I did and helped to sustain the commitment I made to this kind of speculative writing, crazy as it still may seem to you. You both rolled with the crazies and for that you have my everlasting love and deepest thanks.

Introduction
WRITING STORIES OF *MAKE-BELIZE*

Writing is always inseparable from becoming, always incomplete, always in the midst of being formed.
—Gilles Deleuze, *Literature and Life*

It matters what matters we use to think other matters with; it matters what stories we tell to tell other stories with; it matters what knots knot knots, what thoughts think thoughts, what descriptions describe description, what ties tie ties. It matters what stories make worlds, what worlds make stories.
—Donna Haraway, *Staying with the Trouble: Making Kin in the Chthulucene*

A rhizome has no beginning or end; it is always in the middle, between things, interbeing, intermezzo. The tree imposes the verb "to be," but the fabric of the rhizome is the conjunction, and . . . and . . . and.
—Gilles Deleuze and Félix Guattari, *A Thousand Plateaus*

Swirling

Once upon a time a giant Flood devastated the coastal village of Wallaceville, Belize, issuing forth what Miss Grace calls the *beast-time*. And the beast-time ushered in another Flood: waves of tourists, madly buying into new life experiences of that seductive and unsettling azure blue Caribbean Sea. That was around the time a local character named Twitch mysteriously died and Twitch's cousin, a seasoned local Creole named Richie, began to invent twisted stories of adventure for tourists with a wry sense of enchantment and the absurd. And it was just about then that the Belize government drew up plans for a Raelian[1] tourist welcome center. And around the time a crazy white man named Mr. Pete suddenly appeared and began maniacally to rake the village beach for days, agitating local life and gossip. And then he disappeared without word or trace. And that was in the time US money fell from the sky *like mana from heaven*, Miss June said. And that's about the time a mystery ship sailed silently into Wallaceville Bay. And a Belikin beer coaster helped to materialize a wild male strip-show beach party. And that's when a Creole woman named Parca became seduced by her number picks while playing Boledo, the Belize national lottery game, and winning more often than losing while using her numbers to help contemplate future life in a village that had gone *crazy for tourists*. Wallaceville is a make-believe beach town materializing out of this beast-time mix of intensities, images, hunches, strife, enchantment, giddiness, and tactility.[2]

This book is a collection of stories written by way of these odd and restless historical fragments, unfolding as random moments of the ordinary, set against a swirling current of forces that are continuously unsettling the Wallaceville shoreline. Each fragment generates a story that materializes like the heave of curling waves bending to changing tides, strange weather, sudden appearances of seaweed on the beach, cutting waves from the wake of boat traffic, and curious ocean currents fashioned of so many transecting intensities composing Caribbean-Atlantic historical rhythms, vitalities, and patterns that make and remake the ebb and flow of Belizean life (see Benitez-Rojo 1990, 1992; c.f. Sharpe 2016). Like the accumulating jumble of flotsam drifting the Belize coastline as a coiled entanglement of stuff somehow assembling beachfront turmoil, these swirling stories of wonder, luck, and lingering shock, recurring bother, and fresh hilarity energize the place. Each story recombines waves of roiling, excessive telling and possibility that collectively move things and change them across an orbit of a narrative *make-Belize*.

The make-Belize is instantiated in stories that are happenings, contingent re-combinations of images, events, networks, sensibilities, and situations that constitute the unfinishedness of life in Wallaceville and so conjure a Paradise impossible to fix and secure, an unsteady Paradise that stumbles beyond measure or cadence and so is always in the process of rendering place and time sensational in its efforts to make the beast-time resonate sensibly.[3] Yet no efforts of Paradise-production can freeze time or space or transform its forces except through new forces, new compositional energies and inventions that are make-Belize.

But Miss Grace is only one of a growing group of aging Wallaceville locals[4] and old expat locals[5] who today feel that their "time is at hand." Something *crazy crazy* is happening that is deeply felt as both menace and promise. Some "get it." Others don't. For those who don't, it's "get in, take advantage, or get out and get lost." Belize time has entered into what Miss Grace calls the beast-time. And the beast-time took on a vital materializing force and somber strength in Wallaceville with the great Flood of 2001. When the earth cracked open, the beast issued forth, Miss Grace says, and life was sucked out and under with a deafening racket. Disaster. Chaos. The atmosphere was suddenly electric when life fractured in dark sulfurous seizures riding a ruthless sea surge. And the village disappeared. Wallaceville. Gone in a day and a night. Swamped by fetid water then swept away or sucked into the ground.

It is not lost on those who still live there today that this devastating flood struck near the end of a very successful and promising 2001 tourism season that saw unprecedented growth of local tourism infrastructure and a surge of tourists with deep pockets and curious desires for this place they called Paradise. Wiping almost the entire village off the earth, the Flood left little in its wake but difficult questions, like *What else? What next? Why us?* Almost everything was destroyed; the rest was pushed over, under, sideways, down. The traumatizing event gave everyone the *jittas* and sent a nervous shock through the village and the country that is still deeply felt today.

The Flood was a *bad sign*, and it stirred up more bad signs. *It change up everyt'ing*, Mr. Richie said. First, the flood surge, a sure sign of the beast. The staggering chaos of an annihilating disaster that summoned another un-worlding wave, a remaking that assembled as a shocking, uncontrollable surge of tourists with money, with unrealistic ideas about life in the tropics, with odd practices, fashions, diets, and focus. A new flood of strange tourists, and all of a sudden, a stranger, doubtful nature and a nervous economy conjuring some fractured tropical Paradise. *Apocalypse*, Miss Grace whispers. Beast-time energies unsettled things fast and hard. This book marks a time of make-Belize in Wallaceville when the place went *crazy crazy*, for the tectonic shock of Caribbean tourism pumping up a post-Flood Paradise-comfort-zone, summoning beast-time demons and wonderful things.

This book is an attempt to craft ethnographic writing that creates conditions for opening the make-Belize out of those shades of experience that shape the nervous edges, the potentializing forces, of world-making, the non-linear lived duration of experience generating a "what-else" of things "in-action" (Manning 2016: 16–18). Today Wallaceville is a place and time that composes itself in incommensurate registers, circulations, expressions, and publics as beast-time *somet'ings*. These are made in the swirling eddies of dangerous feelings, fierce actions, offbeat moments, strange objects, eccentric ecology, temperamental elation, and touchy beach sensations that pull things into some alignment to become nervously generative of *somet'ing*. Wallacevillians have a strong sense that they are into *somet'ing*, that *somet'ing* is happening, *somet'ing crazy crazy* is materializing that they call the beast-time. This book tracks this unsettling, unending ebb and flow movement—local connections of one unpredictable event pushing and pulling against the next that, in swirling assemblings, make up Wallacevillian experience now.

Stories of beast-time *somet'ings* make Belize, materially. The make-Belize stories that I write are transitive compositions of beast-time make-believe. In kinship with Haraway's "speculative fabulations" (2016: 10), they are transformative because the make-believe moves Belizeans beyond the limits of their own possibility of giving themselves over to the concrete experience and circumstances of other instances of life, event, sensation, and matter, now especially focused on the dangerous prosperity promises of an international tourism Paradise settling in on a compromised marine ecology. It is the transitive potential of the beast-time make-believe, in the forces of movement, to move people and things through encounters of contingent entanglements with countless other beings, things, temporalities, sensibilities, and worlds that activate the transformative potential of make-Belize.

Make-Belize story arrangements exceed the worlds they fashion because "every arrangement installs its own possible derangements and rearrangements" (Povinelli 2014). As in, for every major normatively arranged tourist experience, for example a boat tour that stories an adventurous trip to the Belize barrier reef in attractively packaged images of reproducible tropical adventure, there are minor forces, things moving *otherwise* that course through the experience as other possibilities, like the threat of a shipwreck, a storm's menacing darkness, the fear of getting lost or being abandoned or abducted, the seductive pull of a drinking party, the alarm of a police drug raid, a sad death by drowning, each conjuring a nervous variation, a difference, in relation to repeated and established boat tour enjoyment. This *otherwise* is open to flux and rearrangement, conjuring stories and pulling at feelings that are not controlled by pre-existing, normative structures of tourist image and theme. The otherwise is a metastable derangement of things that conjures innumerable other things that are happening but that remain unfolded and undefined by the established tourism story line and so carry forces into things, wrapping around them, intensifying twists of things that coil like multiple historical-material currents that roll and recombine because of their excessive movement (see Ochoa 2017: 180–81). The otherwise creates a "cut" in the event through which new fields of relation can be fashioned, a composing process that opens experience to new variation, resonance, and expression (Manning 2016: 18–23).

As such, make-Belize stories are always unfinished composings, fabulations, always in a process of becoming more than one thing, theme, or experience. They are potentializing machines that recast the field of potential, open it to contrast, to a change in direction and quality that make a difference felt and so act as incitements of change. This is a book of such make-Belize swirls of incitement, actively generative stories that share their creative energies with the fields from which they emerge, in a state of emergence that makes them powerful potentializing forces of realization. This is the *crazy crazy* that Miss Grace speaks of: the shifty, fugitive, devious, undiscriminating, unstable, erratic, dream-like, eccentric forces that animate life in Wallaceville, the power of which is resonant in everyday sensibilities, emergent vitalities, and immanent possibilities.

Make-Belize

If someone were to ask for a graphic picture of what Belize is like in its Caribbean setting, I would refer them to Benitez-Rojo's (1992: 4) evocative depiction of the Caribbean invented through his figure of a sea-sky:

> **Sea:** aquatic . . . the natural and indispensable realm of marine currents, of waves, of folds and double-folds, of fluidity and sinuosity . . . a chaos that returns. A detour without a purpose, a continual flow of paradoxes; [the Caribbean] is a feedback machine with asymmetrical workings, like the sea, the wind, the clouds, the uncanny novel, the food chain, the music of Malaya, Godel's theorem and fractal mathematics. (Benitez-Rojo 1992: 11)

> **Sky:** the spiral chaos of the Milky Way, the unpredictable flux of transformative plasma that spins calmly in our globe's firmament, that sketches in an "other" shape that keeps changing, with some objects born to light while others disappear into the womb of darkness: change, transit, return, fluxes of sidereal matter. (Benitez-Rojo 1992: 10)

Thinking with this double active figuring of churning sea and swirling firmament about how to make-Belize as an "other" shape that keeps changing because it is perpetually in transition and flux, invites us to speculate on how to compose the present moment in Wallaceville. To make-Belize is to think with and lean into this figure of the Caribbean sea-sky as a multidimensional, push-and-pull material relationality (Benitez-Rojo 1992: 11). As such, Wallaceville is an unstable connection of flowing liquescence and star dust, neither completely aquatic, celestial, or terrestrial, animal, mineral, or vegetable; too much connection to be any one thing, the make-Belize is a speculative "yet-to-come" *somet'ing* that is an indeterminate, transformative, and durative "becoming-with" (Haraway 2016: 10–12).

Make-Belize writing means understanding these processes of attraction and relation, imagining my way into a power fraught beast-time history-in-the-making, still indeterminate but moving *otherwise* into or as *somet'ing*. To do this means imagining Belize as aquatic-terrestrial-celestial. Like the Caribbean more generally, Belize is made up of the indispensable swirls of marine currents and star signs, of iridescent wave formations and folds of fluidity, coral build-up, celestial gravitational push-and-pull, and openly indeterminate, polytemporal swirls of materials, people, languages, temporalities, and histories.

There is a promising liveliness in make-Belize fabulations that unsettle the stories of commercial and social forces of some gloriously luminescent floating European conquest (past and present) that relies on controlling the powers of both sky and sea. Colonizer England, Spain, and Portugal, their once-upon-a-time ships navigating Belizean waters like death-stars of floating menace and light, attraction and repulsion, centripetal and centrifugal forces, pulling into ports of call seduced by the gravitational pull of exploration, extraction, expropriation, and then the centrifugal push home, cargo holds full: a wave-making movement of commerce, dangerous and dense.

Colonizer Europe blasted its energies into space in waves and folds of creation, newly formed elements and molecules that floated off to seed the Caribbean for new generations of colonial satellites to form and grow arborescent. Today, postcolonial commercial efforts rely on Belizean nature, bodies, and culture as commodity attractions, yet again seducing white bodies to Caribbean shores with their own "seeding" efforts of leisure-class social reproduction, mythical tropical life, and image growth of all kinds. But my efforts to make-Belize means pushing past these seedy Euro-exceptionalist assumptions about what counts as a proper form of historical fact-telling, the purpose of which is to participate in projects of decolonizing the past, to critique the inequalities of the present, which includes the assurances of the new financial rule of state-sanctioned, globalized tourism in Belize. This planetary figure of a colonizing, penetrating, phallic-formed cosmos is instrumental in describing the local historical legacies of *anthropo-capitalocene* exceptionalism in beast-time Belize. But to make-Belize in the beast-time means activating storytelling differently than we ethnographers have in the past, to tell stories in an *otherwise* fashion (Biehl and Locke 2017: 5–11; Haraway 2016: 5; McLean 2017a; Ness 2016: 3–40; Pandian and McLean 2017b: 20–21; Stewart 2007).

Staying with the Make-Belize

Many Wallacevillians suspect that the phrase make-Belize was invented by some American advertising agency hired by the Belize Tourism Board in early 2001, notably and ominously just before the Flood.[6] Regardless, it became a pithy sound bite, among others, each bit of language instrumental to an official state-inspired tourism campaign directed at international tourism industry leaders with the hope of encouraging them to sell Belize as a major ecotourism destination. The Belize Tourism Board developed international tourist campaigns using this kind of language to promote the image of Belize as an enchanted, eco-friendly, tropical Paradise, a land of unsurpassed pristine nature, adventurous history, and friendly, attractive culture, the perfect affordable, safe, Caribbean dreamworld escape, a land of make-Belize. Industry leaders, visitors, and Belizeans alike were encouraged to activate generic "make believe"/"make-Belize" alignments that were meant to anchor the powerful Paradise tropicalization stories that endorsed the big picture of every Belizean tourist experience, thereby enacting the sensations of Paradise as a unique self-realization in order to bankroll the nation's future economic success and happiness.

The tourist industry story of make-Belize is a powerful seduction, responsible for putting Belize on the international tourism map as a desired destination, a Caribbean, eco-friendly, tropical experience worth the purchase. The practical and imaginative efforts of the Belize government and the international tourism industry to pin down a tropical Paradise, as well as the capital-financializing logics of incentivized resort, cruise ship, and retirement resort development and management schemes necessary to administer it as a profitable destination, have been very successful, despite the economic fluctuations and interruptions in international leisure markets and housing markets that dramatize the contingency of tourism capitalization projects overall. Efforts such as these are endorsed by much of the applied work being done in schools of tourism, leisure, and hotel management; together they encourage different kinds of "positive tourism" development and management, economic success, and profit.[7] But the positive tourism development-management stories they tell are mired in lands, cultures, and bodies already controlled by the commodity logics of tightly regulated capital and by the moral economies that are meant to maximize the profits of Belize tourism for local elites and international leisure consortiums in the Anthropocene: less expenditure, more efficiency, progress, meaningful product—so more success, profit, and pleasure all around.[8]

This is system-based tourism study. It builds on the understanding of a contemporary Belize economy and society and an ideology of progress and growth for its success. Since its independence in 1981, Belize has invested heavily in tourism as one of the key economic and cultural engines for national development. Belize is sometimes said to be a nation that "skipped modernity" (Sutherland 1998: 3, 4–9), a young country that took a bad bounce into post-Cold War, neoliberal economic growth and development to become a tourist state located on the "roughed up" edges of Empire (Hardt and Negri 2000). Since the 1980s Belize has become linked globally through the introduction of new privatized media forms with the latest mass communications and fiber optics technologies and, not unusually, there are cyber cafes almost everywhere.[9] Almost everyone has a cell phone, Smart or Belize Telemedia's DigiCell, and cable TV, and WiFi. This has allowed Belize to transform its banking and telecommunication systems, its national image production, its propaganda machinery (now linked closely to the Belize Tourism Board), its tourism industry, and everyday community social life through Facebook interfaces, Instachat, WhatsApp, and the "twitter-sphere." It is through these new finance and image-making models underwritten by new social media platforms that tourism is sold off-shore to future tourists and to Belizeans themselves. A 2005 Love FM radio and television advertising campaign made Belizeans aware of their connections to the tourism industry and the future of tourism in their lives. Each segment ended with the telling message: "Tourism is for you, tourism is for me, tourism is for all of us. It is our future, get involved." Today, the Belize government and its agencies advertise statistics that boast a quarter of working Belizeans are directly involved in tourism for their livelihoods.

Historically, Belize never did develop a base of industrial or agricultural mass production as an engine of economic and social success. Until the late 1970s Belize was a remote and internally disconnected British colony without the modern structures of liberal democratic government. It had a "colonial economy based on import and export trade and import-substitution agriculture" (Sutherland 1998: 3). Until it became an independent nation state, Belize did not have its own complete written history, no modern public system of education, and its diverse ethnic population was not recognized officially as it is today, as a unique "multicultural mosaic." Many big changes have occurred since Belize gained independence, and since then Belize has grown under the strong influence of "transnational movements and ideas such as environmentalism, liberalization of the economy, democracy, international tourism, and the international drug trade" (Sutherland 1998: 3).

To this list, I can add the dramatic expansion of the NGO sector, the new offshore banking industry, and the transformation of Belize into a site for global money laundering, the burgeoning evangelical Christian church movements, and the internationalization of local corruption and crime (Duffy 2000). But it is through the dramatic and recent development of international tourism, through the cultural logics and economic supports of the World Bank, the American Development Bank, "Chinese money," "Lebanese money," "Coca Cola money," "Mennonite money," possibly Raelian money, and the money of several wealthy individuals who dropped cash into the country through various investment schemes and opportunities (famously and infamously big rollers like Sir Michael Ashcroft, Francis Ford Coppola, and John McAfee), and a long list of donor interests through which Belize realizes its future and invents its past. It is most forcefully through the introduction of transnational tourism industry (hotels, leisure and fashion industry, sports adventure, nature and culture) that Belize has moved from an economic backwater in the region, and in the world, to a transnational nation tapped into and moving in the gravitational orbit and flows of the global forces of money, leisure, and entertainment, all part of the spectral phantasmagoria of neoliberal capitalism, "Belizean style." It is through an official plan for economic and cultural revitalizations seen through a contemporary ecotourism prism that Belize feels much of its future depends and that deepens the characterization of the *anthropos* as the geological epoch of human-dominated capital entanglements.

You betta Belize it: the all-purpose advertising slogan, along with others, *Make Belize, Belizability, Un-Belizable*, and *Belize it or not*, attach to Belize's "can do," positive, public face and create an official discourse about the promise the nation makes to the world and to itself. Advertised as *Mother Nature's Best Kept Secret*, Belize has done everything it can to cash in on a theme of an international ecotourism Paradise filled with an "awe-inspiring nature," a scientifically managed "exotic prehistory," and a remarkable variety of tours and accommodations (from the latest in full-service, air conditioned luxury to cruise ship mass tourism, to a dwindling variety of cheaper, locally owned and operated services) to make exploring Belize "fun, safe, and exciting." There is a Belize experience for everyone's pocketbook and the *Lonely Planet Guide to Belize* and the *Rough Guide to Belize*, among several other comprehensive guidebooks of Belize, list almost all of them. Belize, like the international tourism industry more generally, has become expert at producing glorious images of itself alongside the construction of monuments, memorials, galleries, "houses of culture," and museums, which act as the containers and vehicles that embody, encode, and distribute the nation state's official heroic, anticolonial master narrative about its rich culture, art, nature, history, and industry, all transformed into a "product" to be bought and sold on the global market. This imagery also works to fashion and define national citizenship as it promotes the official iconography and messages of what it means to be Belizean.

Ironically, all of this has become a seduction engine for waves of the world's white retirement crowd and for those who seek long-term "come and go" arrangements through condo timeshare or private property purchases. Lan Sluder, who writes the tour guidebook of Belize for Fodor, also sells the "bible" of retirement guidebooks for those thinking of retiring to Belize.[10] For the most part retirees and the "nearly-retired" are the "money people," and in Wallaceville in particular there has been a concerted effort to attract them with a local, sustainable ecotourism narrative of laid-back living in a tropical seaside Paradise, close to nature, in the latest condo or new subdivision, accommodations complete with infinity pools and sophisticated security systems, marine toys, and with fun-loving, industrious, and peaceful locals as your labor and neighbors.

While it is images of prehistory, pristine but vulnerable nature, relaxed Caribbean culture, and friendly natives that have been the main attractions for national tourist development and transnational ecological and archaeological interests, the Belize government is now involved in a process of transforming its own population into attractions worthy of national holidays, holiday adventures, and long-term expensive expat Paradise hideaway purchases. Belize has now recognized the great international interest in ethnic and cultural tourism and has begun the national process of inventing cultural traditions, art and music, and ethnic celebrations and holidays for both the global tourist and local Belizean markets (see Holmes 2010; Roessingh and Bras 2003). As Sutherland explains: "There is now a Belize cuisine (before people just ate stew chicken, rice and beans) [see Wilk 2006], a Belizean flag and national anthem, Belizean ethnic goods, Belize beer and Rum, Punta Rock (Belize's own music [along with Boom and Chime]), and Belize beauty pageants . . ." (1998: 186). Belize is available for consumption and every site of consumption promotes local knowledge, national and regional development plans, ethnicities, and cultures as the metonyms of national identity fostering points of identification, common anticolonial struggle, and a unity of Belizean purpose and spirit (see Medina 2003).

Wired into global information flows, Belizeans, who just some fifty years ago lacked more than a passing knowledge of, or local interest in, each other, now find themselves in instant contact with the world. That world appears at their doorstep with new commodity seductions, money, and refugees, get rich schemes, populations, attitudes, diseases, corruptions, debt, fashion, enjoyments, and lifestyles. Today, Belize is located on one of the shifting fault lines of a de-territorialized and transnational global empire, where local worlds are caught up in new fantasies of becoming that in turn act as the catalysts for new life in Belize (see Hardt and Negri 2000; Piot 2010). Such is the context for the critical tourism studies stories of Belize and beyond, set in the context of Caribbean financializations, addressing issues of media, gender, ethnicity, food, health, history, and the like (see Wilk 2006, 2002, 1995, 1994, 1993).

My interest in Belize tourism is not easily captured by the applied tourism work that addresses these themes, even as some of it touches on the turmoil of tourism development and management projects. But neither are my interests completely captured by the work in critical tourism studies with its strong social, economic, and political critiques of the international tourism industry, or with its efforts to take up critically the racial, gendered, sex, and class politics of local tourism development and encounter, or with its critique of tourist representations, or with the critical study of neo-liberal social and cultural constructions of life in a rapidly changing and "developing" nation that relies heavily on tourism today. Critical analysis of these and other tourism-related subjects may be helpful, but the work is mostly burdened by the assumptions of some high ground of strong evaluative critique about tourism, tourists, the industry, and its futures. Here the hope is to recognize and critique tourism logic and practices and identify or even stop bad things from happening and so make some imagined future safer, improved, and more self-aware for coming generations of tourists and the toured. As Biehl and Locke (2017: x) remind us, "the flat realism that comes with the standard practices of contextualization and historicization" is not enough to find ways into arresting encounters, images, and forces of life in order to consider the uncertain and the otherwise incommensurate agitations that open into what Tsing calls "the middle of things" (2015: 251–77). Relying on critical analysis may simply echo the drab determinisms that orchestrate much of tourism theory and social theory more generally. Here, too, tourism, as an event, is held in place and its potential tunes toward reactivity. This is the work of interpretive judgment, not affirmation. By affirmation I do not mean that the work of thought is comfortable and reassuring. By affirmation I mean that "its work is to invent conditions for new ways of activating the threshold of experience, new ways of experimenting in the complexity of what does not easily fold into [position in a field or code] . . . Where negation remains so certain of the stakes of the encounter, affirmation delights in the creativity of *what else* that encounters could do" (Manning 2016: 202). Speculative fabulation is affirmative. It invents and does "its work at the limit where *what if?* Becomes *what else?*" (Manning 2016: 202). It is the *adding to* rather than the *adding up* that interests me here.

There is a long and productive history in the critical studies of tourism dealing with questions of encounter as self-other relationships, spatialized embodiments, identity politics, gender politics, racialization, and sexual relations (see Brennan 2004; Castaneda 1996; Frohlick 2013, 2016; Frohlick and Harrison 2008; Kempadoo 2004; Picard and Di Giovine 2014; Salazar 2010; Salazar and Graburn 2014; Simoni 2016; Shepard 2018; Smith 1989; Thomas 2014; Urry 1990). This literature consists mostly of rich and detailed ethnography framed through one or another social and cultural constructivist approaches to tourism and I make reference in the short list above to some of the very best of this material in order to acknowledge its important goals while gesturing to how the work as a whole still relies on a dialectics of self/other and subject-object relations and gesturing to the questions of encountering and representing otherness in the context of tourist spectacle consumption and the world as exhibition (Bærenholdt, Haldrup, Larsen, and Urry 2004; Little 1991; Mitchell 1988; Urry 1990).

A long look through these works gives you all of the significant references for critical tourism studies since its inception. But what is especially important about this most recent critical ethnographic work on tourism is how it focuses on the politics and poetics of representation or on the social, gendered, racial, economic, and power dynamics (colonial and postcolonial) that underwrite the uses, strategic or otherwise, of cultural images, inequalities, imaginaries and practices (both every day and exceptional) for personal enrichment, entertainment, and exhibitionary purposes (see Bruner 2004; Crick 1994; Graburn 1983, 2001; MacCannell 1976, 1992).

Regardless of how critical and valuable this writing and thinking is to tourism studies in Belize, its analytical focus is to adopt useful concepts and to apply them in order to build a system or a metaphor to better understand the tourism problem at hand: getting the "problem" of the tourist situation, the tourist, the industry, or touristic encounters right. But "getting it right" means, as Massumi puts it, developing analysis that

> . . . disavows its own inventiveness as much as possible. Because it sees itself as uncovering something it claims was hidden or as debunking something it desires to subtract from the world, it clings to a basically descriptive and justificatory modus operandi. However strenuously it might debunk concepts like "representation," it carries on as if it mirrored something outside itself with which it had no complicity, no unmediated processual involvement. And thus could justifiably oppose. (Massumi 2002a: 12)

While both critical and applied approaches to tourism may be useful and evaluative procedures for tourism studies, they are not mine.

Contemporary shifts in tourism studies have introduced a new level of concern with the spasmodic effects of local and global flows of capital, information, people, and culture (see Colman and Crang 2002). Acknowledging such effects, attention in tourism studies is now turning to how tourist encounters are performed, enacted, and embodied under the exhilarating and anxious conditions of transnational cultural mobilities and sensibilities (Franklin and Crang 2001: 17–19; Fullagar 2001). In their editorial for the journal *Tourist Studies*, in which they describe the contemporary troubles with tourism research, Franklin and Crang (2001: 17) comment on the "welcome attention" now being paid to questions of tourist enactments and the "move beyond the study of representation toward seeing tourism as a system of presencing and performance."

Favoring a non-representationalist, assemblages, productivist approach (for examples see Crouch 2010; Ness 2016; Saldanha 2007; Tucker 2017) instead of a dialectical, representationalist, reactive one, means that my writing and thinking are always in a productive tension with critical and applied tourism studies, whatever their approaches are to creativity, invention, and relatedness (See especially Bruner 2005; Frohlick 2016; Harrison 2003; Hutnyk 1996; Kaplan 1996; Kirshenblatt-Gimblett 1998; Picard 1996; Rojek and Urry 1997; Salazar 2010, 2013; Scott and Selwyn 2010; Urry

1990). Nevertheless, my writing, my storytelling, is meant to provisionally grasp the embodiments of rapidly changing material-semiotic encounters through tourism, in hopes of evoking some idea of what Franklin and Crang (2001: 17) mean when they go on to say that: "Tourism is a productive system that fuses discourse, materiality, and practice." Unpacking this phrase has helped me develop questions about the way tourism in Belize may be thought of, through an affirmative, vital materialism generated as ecologies of the insensible. By this I mean fields of sensation and assemblages of forces are emergent properties of human-nonhuman, material-semiotic liveliness, forces that impel new tendencies for life, new forms of fascination, vitality, attention, and distraction: "sense as matter forming, as cohabitation," forming "new practices of sensation and new sensibilities" through such unruly and intractable issues as encounters in tourism development, ecological change, and social life reanimations (Yusoff 2013: 208). This has more to do with opening up possibilities than forging judgments while enforcing analytical schemes. Such writing practices disregard other potential sources of creative emergence by bracketing whatever is not already given according to the strictures of representation itself. This limits things and reduces the unexpected, as Deleuze says, to the invention of forms not the capturing of forces (2003: 48).

I came face to face with having to think about the enactments of encounter in tourism studies while conducting my Belize fieldwork, in the awkward but always intense nature of configuring imaginatively the milieu in which a tourist and a local, a scene, a performance, a culture, a dance, a thing, an animal, a movement, a disaster, an elation, an ethnographer, or a nature meet and what is activated in the contact. Ethnographic-tourist encounters examined by tourism scholars are almost always taken up in the context of "the one and the other," subject and objects fashioned in that space of the "meeting ground" of "hosts and guests" while unraveling the political, social, economic, sexual, gendering dynamics of the engagement (see MacCannell 1992; Smith 1989). Instead, to think the multiple activations and practices of which the encounter experience is composed is to think with encounters as emergent moments of chaos, chance, insensible attunements of time before they are named as "encounters" proper. "What else could be at stake in the encounter if it weren't organized around the certainty of knowing? What might become thinkable if knowledge weren't so tied to an account of subject-driven agency? And, what else might value look like if it weren't framed by judgment" (Manning 2016: x)?

This book, therefore, through the stories it tells, engages the anthropology of tourism with a literature that it mostly doesn't think with, namely critical post-humanist work in ontogenesis, concerned with how encounters "matter," and how matter is thought and constituted through vibrant entanglements, refrains, knots, and figures of human and nonhuman bodies, affects, objects, and practices. Attending to tourism studies thusly reorients thinking around questions of relationality, about the resonances of material-semiotic forces co-implicated in what bodies can "do" and how "matter" "acts" rather than a concern with what "is" a body or the agentic meaning of experience when considering the relational processes of encounter. Here, I am interested in how tourism activates potentialities in bodies to be otherwise, to generate certain kinds of Paradise natures, mutations, and affects as insensible natures, as the agitation or provocation, and curiosity and desire that draws over the work of intelligibility in acts of encounter. So the insensible draws its energy as an agitation in movement what Grosz calls "nicks in time" or Barad thinks of as "quantum cuts" or Deleuze and Guattari call "virtual ecologies" or Massumi calls a "rhythm without regularity." These are focused indeterminacies, in the example of Caribbean Belize sea-sky swirlings, and attention to them might possibly energize thinking through encounters in tourism as indeterminate processes that remain otherwise, incommensurate with given forms of knowledge production that anchor the study of encounters in tourism and tourism studies.

What would it be like to think of the indeterminacies of encounters as an ecology of insensible practices (Yusoff 2013), a movement of thought, oscillating forces of becoming active, activating new registers of "becoming with," co-constitutions that enact a contingent durability conjuring an ethical concern as a political yield? And how does the work of encounter include our own encounters too, with writing the limit, as Blanchot (1997) says? Encounter writing, writing at the limit, at the edges of what can be thought and felt, the unspeakable "more than": how might it be possible to write as an ethical practice of making a difference? How does such writing activate a poetics of difference marked by how our senses might attune to artfulness and storytelling in other modalities, as stories told differently and in registers not eclipsed by an ideology or a structure, by binary thinking, or by describing some moment in time that keeps things on track and in time through the usual sequential writing practices of clarity, disavowal, and debunking? By binary thinking I mean the usual "us and them" "host-guest" descriptions of encounters that still ground tourism studies, still caught up in the referential logistics of dialectical thinking and practices of description that do not begin to describe the generative forces of the *what else* and the *otherwise*, work that reaches beyond the security of evaluation and solutions to what is at the heart of "speculative fabulation" (Haraway 2016) to create an "untimely" mode of active inquiry into the *what else*.

So an important activation experiment that makes up a central theme of this book is how to take a concept like the "moment of encounter" in tourism and imagine actual encounters as difference generators and thereby, inspired by the curiosity of what is the otherwise or the insensible of encounters, draw a different attention to the nonlinear lived duration of encounter experiences, attending through our own experiments in writing to cocomposing moments as the creative generative intimacies that are the contingent and unfolding enactments of encounters themselves. Working along these lines generates my interest in the beast-time make-Belize. My politics and ethics here are anchored in questions about how to relate, how to write, how to "sense beyond security" (Manning 2007: 134–61), and how to become touched with that which is beyond you and me and yet is cocomposing, as things forming beyond the exclusionary tactics of working with representations of encounters and so with the term "encounter" as already framed and anchored in thought and practice. This is storytelling that works the "as yet unthought" in the act of its becoming: an act of make-Belize.

Through the processes of make-Belize, invention things modulate, affect, and transform each other. The materializing metamorphosis of acts of Wallaceville make-Belize are the processes that transform one half-formed historical thing into networks of other half-formed things as if by some crazy conjuring trick right before everyone's eyes: an apocalypse out of chaos, the narrowing specification of chaos from a particular point of disorder and unpredictability, a profusion of forces gathering and intensifying, a materiality mustering a measure and a rhythm out of nature without norm, into *somet'ing* until the next *some'ting* begins to happen. Attending to these ebb and flow processes as newly forming rhythms of life, a temporality with its newly textured and unsettling sensations and new forces and energies, I encounter the beast-time in Wallaceville.

My project is an attempt to conjure beast-time make-Belize worlds out of the world that we all share, if unequally, to detail what Tsing calls "the possibility of life in capitalist ruins" (2015). Conjuring inspires new ways of telling stories beyond the strictures of the master narratives of modern economic rationality, progress, and growth or of its critique. But telling make-Belize stories is not an act that neglects the horrors, death, and destruction of Belize and the planet or the urgency of "staying with the trouble," as Haraway (2016) puts it, and attending to the ecological mess we all are living. There is no intention of replacing this story by slipping into some idealist, utopian, or romantic mode. The aim here is to craft a make-Belize as an otherwise, for better or for worse, that manages to live despite capitalism's ruinous forces as a logic-action machine. There is always "more there, there," as Nigel Thrift (2004: 58) says. Tending to the muffled or stumbling rhythms and registers of becoming, to the entangled material-semiotic-affective dimensions of life popping in and out of possibility, widening the potential number of relational connections that things can enter into, that can increase the density of transformational possibilities in the act of their composition the act of make-Belize writing is a conjuring, an improvising with worlds in the making, unfinished, becoming.

Hope and Storytelling in the Beast-Time of Make-Belize

> Real problems must be given time.
> —Henri Bergson, *The Creative Mind: An Introduction to Metaphysics*

> Full-bored ethnographic writing tries to let the otherwise break through, to keep it alive. To tend it.
> —Kathleen Stewart, *Crumpled Paper Boat*

The temporal grounding of the beast-time in Wallaceville is the different capacities of turbulent vibration (unliveable power) to link bodies and sensations into forces of becoming. Make-Belize stories are acts of extracting something of these sea-sky swirling forces of the ebb and flow of events, stories, and new sensations that are activated as incipient social flows, material relays, attunements, images, and life practices. Waves of ebbing sensation compress things, and chaotic forces condense into potentializing patterns of polyrhythms: acts of beast-time shaping and creating out of forces of swirling vibration. Waves of flowing expansion dissolve movements of rhythm, blurring them back into the resonances of sensation. The relation between ebb and flow is precisely definable in terms of rhythm and current, not as a mathematically measurable and finalized form but as a duration-movement, uncountable because it is always in process, an open-ended movement of compression and expansion, the common ground of the durative make-Belize or what Deleuze, while working with Francis Bacon's art-making process, calls "the power of the future" (2003: 54). Make-Belize storytelling is fabulation, a mode of attending to the rhythms and textures of things as they differentiate and compose vibratory forces that constitute atmospheres and worldings as future-making acts. Deleuze says that this kind of writing is an art practice—in my case writing make-Belize stories—when it aims to capture the "force of time," by which he means opening sensation to the force of the future, sensing time, not in order to control duration, *chronos*, but to live it as one can, *kairos*, even if that means becoming-different (2003: 52–54), becoming-monster.

My storytelling is "an activist, occurrant art" of fabulation (Massumi 2002a: 17), a process of composing that, as I said earlier, adds to the making of stories even if those stories never add up. By that I mean that my storytelling is an act of "telling together" that extracts something of the strains of uncertainty and vulnerability of Belize Paradise event-fragments captured through the very force of, and encounter with, the writing. This composing practice narrows my focus onto these particular event-fragments enough that a certain *somet'ing* begins to take form in the rush of Wallaceville nervous intensity and yet still picks up on the imponderable forces of emergent potential that make this crazy Paradise impossible. A Belize vibratory pulsing: a "rhythm without a regularity" but still something durative, time unsettling because it is busy forming out of "too much connection": an act of becoming as emergence, the insensible.

When *somet'ing* takes form as an expression in Wallaceville it instantiates things, it acts as an incitement, rather than representation. Make-Belize *somet'ings* are becomings that Stewart calls "emergent vitalities." As such, they are what Sianne Ngai (2004) calls "bad examples" (see Stewart 2003b). As emergent vitalities, the seven stories I tell are bad example incitements. Stewart goes on to say that bad examples are

> ... not perfect representations of an ideology or structure at work in the real world, but actual sites where forces have gathered to a point of impact to instantiate something. They are not things gathered under the sign of meanings or types but radical singularities with particular texture and density, hybridized mixes of things, flirtations along the outer edges of a phenomenon, or extreme cases that suggest where a trajectory might lead if left unchecked. They are not representations at all but constitutive events at the point of their affective and material emergence. (Stewart 2003b)

The stories that partake of the forces of becoming compose the make-Belize in Wallaceville. Taken together they are meant to unofficially deconstitute the European-based time of "a world picture," as Heidegger called it, or what Timothy Mitchell (1988) calls "the world as exhibition." In Belize today, the political certainty of the phantasmagorical world picture, as its very own "anthropo-*scene*," is underwritten by a story of pristine tropical nature and culture under threat and close to social, emotional, and ecological collapse. Save the planet scenarios run through national and international ecotourism campaigns: "save the reef," "save the jungle," "save the culture," "save the beaches," and "save the manatee, lobster, conk." Belize tourism generally, and volunteer NGO-tourism organizations and some evangelical missionary operation (as we will see in Chapter 2) specifically, thrive on multiple iterations of a conservation ethos that juxtaposes seductive Paradise nature and culture scenes against dire warnings of decay and death that lie just around the corner for Belize if humans are not careful (see Taccone 2019).

But such scenes of salvation instantiate a disquieting local Wallacevillian resignation, "a participatory drag," as one unsettled tourism group leader put it, describing what she found to be a shocking disinterest among old expat and Creole locals in Wallaceville to actively participate in social-ecological salvation projects. Volunteer tourists and the tourism industry more generally find this "attitude" to be disturbing because it seems never to be categorically "reasonable." Yet the chaos of an inchoate future, an otherwise that we cannot yet picture and over which there is no human agency, only desperate and urgent beast-time signs and moves, instantiates a "will to be otherwise," a radical immanence governed by things at hand, "unfinished not merely because our mind has not yet succeeded in categorizing it like scientists now sequence DNA but because in attending to it in a certain way we pull it into being in a way it was not before we did so" (Povinelli 2016: 15, 137). Such semiotic "reasoning" is not a decoding exercise (see Kohn 2013: 15, 50–51; Ness 2016: 3–40). Rather, it is an assemblage and coordination of the habits of things and beings that are ceaselessly "becoming otherwise" in the acts of their assembling formation and coordination.

Everyone in Wallaceville has felt the Beast's presence, movements, and transformative powers. It is through these expressions of the Beast's time: flooding tides, strange weather, an increasing number of hurricanes (while I was writing this as a first draft, 3 August 2016, Hurricane Earl was bearing down with a growl, Belize City square in its menacing sight lines), "nature-culture" extinctions of all kinds, burning jungles, destroyed reefs, potable water shortages, staggering garbage pile ups on shore and at sea, massive, mysterious Sargassum seaweed "invasions," and an alarming increase in population density that has accompanied the influx of new and demanding wealthy lifestyle migrants, cruise ship chaos, new tourism financializations, new international tourism salvation programs in tension with short-term tourist excess, and so an unprecedented pressure on all resources, human and nonhuman (ask the poor sea cucumber). And for those who live in Wallaceville, beast-time signs of chaos, confusion, and destruction are palpable, and they see how such signs initiate "big fix" intervention movements and schemes. The beast-time is upon them and us both, along with all of the unsettling social, economic, and ecological anxieties through which it is fashioned. Crisis looms large. Chaos is a deeply real and haunting presence and it overwhelms the imperfect present. And yet might there be some hope in make-Belize stories or at least an otherwise, experimental time-space conjuring, a speculative futurity, an invention-creation interplay?

Make-Belize as an act of making and remaking the relationalities of life entanglements is a different kind of ethnographic practice. Writing ethnography as lively enactments of "becoming-with" encounters opens risky and uncertain relational extensions that generate make-believe spaces in which diverse bodies of all sorts, human and nonhuman, "succeed one another in mutual movement; hence the continual movement of parts of a body toward parts of another body [not necessarily human or living], across a continuous, uninterrupted space, as if the void was the mediator between two plenitudes" (Sonnier and Donne qtd in Massumi 2002a: 258). This book grows out of this hope for an ethnographic practice that might share in the creation of worlds otherwise, a tenuous and fraught effort to pursue a project of ethical transformation, writing stories as an experiment in attuning one's fugitive feelings "otherwise," something heretofore felt as *potencia*, still unspeakable but in transit, in movement (see Braidotti 2002). Such story-writing practices instantiate passions as the power to act itself, insensible, moving force into matter as potentiality. Where there is potentiality there is the hope of finding new ways of reuniting human purposes with the endless yet precarious vitalities of the world in which all things and beings play at making life liveable, and so co-creating possibility in new forms of fabulating the life of the world (see McLean 2017a).

While writing this ethnography I think again and again about the processes of make-Belize as the work of experimental expression, in the world of relational life enactments in Wallaceville. The expressivity that ties work and world is something more immediate and intimate than ethnographers of tourism are accustomed to acknowledging. Like the magic of storytelling I spoke of earlier, language works with the substance of the world in the way Giordano Bruno, after Lucretius (Nail 2018), reminds us. My story writing engages with transformative potentials immanent in the Belize world at hand.

Affirming the actual has always been imbued with the possibilities of being otherwise—the "always more than" of unruly *potencia*, of things in the act of their composing a reality that is always different from itself—my writing presses closer to the unrest and turmoil that is immanent in the emergence of a discursively knowable Belize world. My writing is meant to stay with the transformative metamorphosis of make-Belize, its disquieting becoming. As such, it is fashioned of its own excessive and transgressive exteriority, a writing that drops me into a potency and peril that can never be mastered but remains excessive of such powerful conscriptions of mastery to act as an unruly capacity, a "minor gesture" that challenges those given orders of meaning, legitimation, and power: unsettling acts of Belizean worldmaking (see Manning 2016).

Provocations

My make-Belize opens several provocations that I hope to address in the following chapters or that I think are necessary vocabulary reminders that reference my unconventional use of language.

1. What is ordinary about how life unfolds in tourist encounters in Belize? Of what processes are unfoldings generated (see Stewart 2007)?

Here I am especially interested in the affective intensities that generate the movement of an unfolding in the context of travel and the play of tourist temporalities consistent with movement and change as they take on expression in Belize.

2. While the expressive economies that configure the temporal dynamics of encounters in Belize are created in terms of tropicalized representations of nature and culture to create nervous Paradise images that buckle under the burden of the work they are expected to do, the collective challenge of this Belize "bad example" is in how it may help point to the specters of imperial tourism capitalisms as systems of capture by attending to their destabilizations and to the contingent articulations and the inconsistent sense-uncertainty of emergent capitalisms and globalist forms.

3. The temporal dynamics of unfolding tourist futures, I suggest, burden the linear sequencing of time and the rationalist logic and binaries through which a future is determined differently from the past and the present. This is enacted in each case through gaps. Mind the gap, Tsing (2004) cautions, those nervous zones of erasure and incomprehensibility. Gaps occur where official and formal tourist-futures projects do not comfortably move or reach so far (or deep) as to change everything according to their plans.

4. Encounters. Is there more to them than what is commonly understood as the act of a tourist and an other, establishing a relationship and a contact zone? I use the concept of the milieu rather than the more recognized concept of the encounter as a "contact zone" in order to distinguish the way the zone is often used as a social description of liminal creativity, as a space of identifiable affects, a sociocultural zone of mobilizations that engages given political and economic infrastructures not from one side or the other (tourist or local) but from their merging. Encounters may include all of this in that they are interstitial and engaged, but they are more than that too. They are more like different edges of an interface touching in a moment and then changing—moving thresholds. As both real and ephemeral fields of movement, a milieu is also a trajectory enacting new infrastructures of difference, new modes of relatedness that affirm, augment, and keep the interface of encounter open to the multiple occasions of its collective presencing. "The trajectory," Deleuze says (1997: 61), "merges not only the subjectivity of those who travel through a milieu, but also with the subjectivity of the milieu itself, insofar as it is reflected in those who travel through it."

5. Affect: there is a tendency in tourism literature to invoke affect largely as a placeholder concept, bracketing more substantive concept work

and the growing concerns with affective thresholds of political agency in an age of neoliberal capitalist ruins and epistemic crises (for an example, see Buda 2015). This poses timely questions, and also finds a corollary in shared ethnographic engagements with that which is otherwise, incommensurate, emergent, or immanent. My hope is that ethnographic engagements in tourism studies will further turn its attention to the ineffable as a vital dimension of social existence or attend to the labors by which the elusively immaterial and ephemeral is made tangibly present to exert material force as the everyday unfolding thresholds of gender, sexuality, class, race, debility, and age. This is to suggest that a power of writing tourism ethnography lies in its abilities to surprise, provoke, and trouble the descriptive habits of systematizing and critical explaining and so push the borderlands of concepts and writing in a way that can destabilize our sense of what constitutes our object of inquiry, our stances toward it, and our ways of communicating it. This is not to argue that the conventional genre claims of tourism ethnography have no merit. And this is not a plea for a new proscription for writing ethnography. It is simply an invitation to all of us writing tourist ethnographies to help invent and share new ways to describe the protean, the as-yet unformed, the unfolding, and the socially fluid.

Notes

1. The Raelians are a science-religious movement. Members believe the "Elohim," an alien race of extraterrestrial super-beings, planted life on Earth 25,000 years ago. They created human life by cloning it. Raelians believe in immortality through cloning. They seek to establish an earth-based embassy as a welcome center for the "Elohim," a sort of tourist visitor center that will also provide the facilities for a super extraterrestrial computer to be used for future cloning purposes. The Raelians consider Belize to be a possible site for the center if their initial plans of building a center in Israel fall through. They have visited Belize on several occasions for high-level talks with the Belize government eager to attract "foreign" investment (see Chapter 2 for more details).
2. There is no actual village of Wallaceville, Belize. Wallaceville is a make-believe place and presence, a composite of several coastal villages and towns heavily invested in tourism that populate the shores of Belize from north to south. So too are the names of the people and events of which this book is composed. They are made up, fictions created of people, events, and experiences that I encountered in Belize. I have moved into the realm of the "all made up make-believe" not only because doing so meets with the kind of literary fictive ethnography I am trying to write but also out of respect for various villages, village councils, and for

the individuals who have requested anonymity for the sake of their privacy, their reputations, and to ward off bad press that may encourage future tourist to avoid these places, people, or Belize altogether. Some of the stories of people and places could be "bad for business as usual," as Mr. Richie once said to me. Wallaceville, too, is a made-up name that incites a relation to Belize history and agency. It fashions a place that is an unfolding present, a fabulated ongoing presence. On 21 June 2007 I came across a document in the Belize National Archives in Belmopan entitled "The Origin Of The Name Belize." It is a record of very early name changes with documentary evidence to support them. It is said that in the seventeenth century a "daring Scotch buccaneer" by the name of Peter Wallace established a small village that soon became known by his name "Wallace." The village was established on a river that soon took on his name as well. There these Scots buccaneers raided Spanish and English ships of their loot only to escape capture by moving effortlessly through the cayes and reefs to disappear up the river that protected them from detection and attack. The name "Belize" is the result of several bad Spanish pronunciations of Wallace's name that are referenced in historical documents. The evidence is found in several letters and reports to British and Spanish authorities. In 1677 there is a Spanish record of Wallace's name in referencing his river settlement, "Balis." By 1705 the Spanish used the name "Bullys." There are 1720 and 1724 documents that named him and his settlement "Bellese" and "Valis." The latter name morphed into "Valiz" by 1783 and "Valix" by 1785 but this time referencing the place and not the person. In 1790 the name changed to "Belize" and from then on, the river, the settlement, and the region became known as "Belize" and Wallace was all but forgotten. In the spirit of name metamorphosis, I deploy *Wallaceville* as yet another creative historical erring that combines "Wallace" and "Belize" into an otherwise imagined place.

3. On the concept of the unfinished, I draw attention to the volume edited by João Biehl and Peter Locke, *Unfinished: The Anthropology of Becoming* (2017). Among the most recent experimental works of ethnography useful in working out the form of this book I am grateful to these authors for helping me develop the concept of the unfinished in the context of Belize and tourism studies. In the same manner the work of the "Crumpled Paper Boat" collective in *Crumpled Paper Boat: Experiments in Ethnographic Writing* (Pandian and McLean, eds., 2017a) was also useful. See also the volume edited by Gretchen Bakke and Marina Peterson, *Between Matter and Method: Encounters in Anthropology and Art* (2017). On the processes of creative writing and imaginative storytelling more generally it is essential to reference the volume out of the Center for Imaginative Ethnography: *A Different Kind of Ethnography: Imaginative Practices and Creative Methodologies*, edited by Denielle Elliott and Dara Culhane (2017).

4. The Wallaceville locals in this book are for the most part Creole. *The National Tour Guide Training Program Manual* (2001), the text that every tour guide working in Belize must now know in order to prep for the big test necessary to earn an official tour guide license and so be able lead tours for visitors anywhere in the country, describes "Creole" in the following way:

> In Belize, the term Creole has been defined as a person of mixed European and African ancestry. Immediately after slavery, the term was also applied, in very few cases, to some locally born whites. Leo Bradley elaborated on the term, writing that, "Creole as used by us, and referring to one of our ethnic groups, means that anyone who has a tinge of African blood, however small or however large." (2001: 215)

This definition is as close to a popular definition of Creole in Belize as there may be in both its historical and contemporary contexts. This definition may answer the question about what Creole means or how it is fashioned essentially, but it doesn't address, in fact it studiously avoids, questions about what Creole, as a designation and a population, *does* when it is deployed as a term and especially in relation to life in Wallaceville today, under the dramatic pressures of change that bear down on this place that are the result of what I am calling the beast-time. For further complications on becoming Creole and local see Johnson (2019) and Wilk (1995).

5. By "expat locals" I mean those white visitors who came to Belize for a holiday or maybe to work on some special project but remained to make the place their home. Locally they are known to each other and to indigenous locals as *come-stay* expats. These are people who now have occupations in the village as professionals. Others once had businesses tied to the tourism industry but are now retired. They sometimes do odd jobs loosely connected to the tourism industry but usually in very informal ways. *Come-stays* have lived in the country for years and are commonly understood to "get it" when it comes to understanding the new tourism industry and what it is doing to Belize communities. They are insulted when confused with the new, much more aggressive, less locally attuned, and entrepreneurial *come-stays* who are the newest residents. They are often retirees who find economic "opportunities" and are sure that they are improving the community with their efforts. The other group are *come-go* expats who, like Canada–Florida "snow birds," spend the winter months in Belize. They usually own property and have been following this come and go practice for dozens of years. These are mostly retired people or those who have professions "at home" that allow them the opportunity to travel like this. Then there are short-stay tourists who may own timeshares or their own property but do not spend more than a couple of months at a time in Belize. Short-term tourists are those that arrive as part of a packaged deal tied to one of the gated communities or resorts or they are the more adventurous do it yourself "ecotourists" (once called backpackers) who spend a short time in Belize "exploring." You find them at local hotels and cabanas owned and operated by local tourism entrepreneurs.

6. There are several tourism advertising slogans that are plays on the word "Belize"; "Make-Belize" is one of them. I will introduce several others later in this chapter and use these slogans for my own purposes in Chapter 1. As for the slogan "make-Belize," many Wallacevillians are not sure which tourism advertising campaign it came from and I cannot trace the term to a single one of them.

7. The range of literature in tourism and leisure studies dedicated to positive tourism or to the "negative and positive effects of tourism on the political, social,

economic and cultural development" of regions around the world is massive. Several presses like Routledge, Channel View, Ashgate, and Elsevier have produced large and successful publication series dedicated to the theme. Altogether there are over three hundred journals devoted to tourism studies, a majority of which are committed to the analysis of tourism planning and "tourism as applied economic progress." The number of these journals grows as the number, impact, and significance of faculties and schools of tourism and leisure studies expand dramatically. A list of references is too large to reproduce here but it is dominated by a focus on applied studies in tourism and of the tourism industry. This was a focus of conversation during the Plenary Panel: "Ethics, Creativity, and Diversity: Understanding and Changing Editing and Publishing in Tourism," Critical Tourism Studies VII Conference, Palma de Mallorca, Spain, June 28, 2017).

8. For an example of the work consultants do for the Belize Tourism Board, the Ministry of Tourism, Civil Aviation and Culture, and the Government of Belize look no further than studies such as "A Social Viability Assessment of Cruise Tourism in Southern Belize" a report (no date) submitted by Seatone Consultants of San Francisco, USA. Seatone conducted a confidential "situation assessment and review" with 28 "stakeholders" in southern Belize in 2010 for the Belize Tourism Board and Ministry of Tourism, Civil Aviation and Culture. The fifty-nine-page report provides a positive analysis detailing how "cruise tourism" (not cruise ship tourism) in southern Belize fits into the Ministry of Tourism's National Sustainable Tourism Master Plan. Seatone Consultants has conducted several other environmental and tourism related projects in Belize (Seatone Consulting 2015).

9. The odd predicament of introducing space-age communication technologies before the thought of developing even the simplest, most modest, and most conventional modern waste management infrastructure or reliable roads, or a hydro grid to support it is more than telling about how priorities are set and put into practice in Belize.

10. See Lan Sluder, *Easy Belize: How to Live, Retire, Work and Buy Property in Belize, the English Speaking, Frost Free Paradise on the Caribbean Coast* (2010); or Bill Gray and Claire Gray, *Belize Retirement Guide: How to Live in the Tropical Paradise on $450.00 a Month*, 4th Edition (1999); or Bob Dhillon and Fred Langan, *Business and Retirement Guide to Belize: The Last Virgin Paradise*, 2nd ed. (2018); or the popular International Living Inc. website, which will send you a free copy of their report, "What You Need to Know about Retiring in Belize."

Chapter 1
"FOR THE TIME IS AT HAND"
BEAST-TIME *SOMET'INGS*

> Blessed is he that readith, and they that hear the words of this prophecy, and keep those things which are written therein: for the time is at hand. . . . Now write what you see, what is and what is to take place hearafter. . . . The words of the first and the last, who dies and came to life.
>
> —Rev. 1:3, 17; 2:8

Un-Belizeable: The Flood

Miss Grace remembers only what flowed through her on the day the beast-time took firm grip in Wallaceville, when the Flood surged and everything was swept away. That's when life turned impossible on the promise of a new beginning. *Di Flood. A naitmyaa*, Grace said. And it ripped the seams out of everyone's dreams of a new millennium Paradise. Belize, late fall, 2001. End of the world? Maybe. Don't die. Chaos but not apocalypse; not yet (Benitez-Rojo 1992: 10). There was warning. Clearer in that moment than she ever was. When the darkness fall. She. Feel. House. Lifting. Floating. Flooding. Head banged. *Laad help mee*, Grace pray. Across the village. Tilting one way. Then another. *Boat in ah stammi see, mi tink*, Grace say. Knee deep in water rushing through her. She with cat-dog-parrot-Bible. The Ark? Like Noah? Not like Noah. Things on the surface of churning water that she want no part of. But like Noah, she whispered, *God say wi had it comin'*.

Yes, there was warning. Count the endless darkening days of heavy wet storm. A tropical low-pressure system pressing in like the full weight of gloom. A choppy sea. Rising. Impossibly wet green sky. Oppressive heat-weight-wind. Then greener darker skies the color of sick and snakes. Geckos chirping alarm. Grace's flamboyant didn't bloom that year. First time ever. The color of the sand shifted to dark. The tiny half-felt earth tremors. The social temperament of the village. The explosion of local thieves. Each an intensity building a network of forces; an emerging entanglement of overall growing edginess. It was nothing Grace could measure or quantify. But it was *somet'ing,* she say. And together this *somet'ing* unfolded as a reckless intensity, sticking to her, incipient, the encroaching threat of something unbidden, but surely felt as material bodily buildup or crack-up. Overcharged signs, an excess of tingling intensities, affective forces forming an event in the act of its becoming, its actualization, an immanence, things feeling like they are charging up, something unpredictable, something like an edgy nervous movement, like the *jittas,* like the *crazy crazies,* Grace say.

Yes, there was warning. The morning the beach dissolved. Impossible, but it disappeared altogether. Gone. No sand. No beach. Hundreds of coconut trees tumbling over naked networks of roots. Worn away by the growing storm surge maybe.

Inhale. Everything floating out, bay to sea, in great pools and strands of stuff. Leaving tiny seaside cliffs black where sand once was.

Exhale. A stench slop sea *kom bak.* Different stuff returning on dangerously high tides. An evacuation of its own sort. Sea breathing out churning eddies of rancid garbage, house parts, bits of coral, dead fish, and sea grass.

Next day. The ground was saturated so now the water just collected, flooding the lowest lying places and rising. The water began to move fast taking great swaths of village: mango trees, gardens, pets, washing machines. Screeching voices in thunderous winds. *Di vais of di Beast. Di blow keep comin', somet'ing crazy wild. Me feel di Beast comin',* Grace recall. She pray. Her Bible turned to the Revelation to John:

> And I stood upon the sand of the sea, and saw a beast rise up out of the sea, having seven heads and ten horns, and upon his horns ten crowns, and upon his heads the name of blasphemy. (Rev. 13:1)

She conjures a world that will be as her Bible say. St. John's apocryphal telling, an intensification of what is real and upon her, the Book of Revelations materializing an unexpected, monstrous realization felt in relief. Shivering, she *mek* her Maya prayer. A blessing for her house, trying to protect it against rising winds and waters, *Henicuchat anapishat anushi.* An Obeah curse too, but the Beast come anyway.

Next day. Evacuation. Emergency. Village pack up over soggy trembling surfaces. Time to get out. The mass exodus to higher ground in the Maya mountains. Everyone go. Now. But Miss Grace. Forgotten. Left behind. Wouldn't leave anyhow. Ride it out. She would. After almost a whole life in her precious Wallaceville, two husbands, two dead, eight children, two dead, twenty-one grandchildren, one dead, three great-grandchildren. Seventy-six years. She ride it out. Grace found an old life-preserver and put it on. And Miss Grace read:

> And I beheld when he had opened the sixth seal, and, lo, there was a great earthquake; and the sun became black as sackcloth of hair, and moon became like blood.... (Rev. 2:12)

Later that night a flash-angry groaning ripped the black-gloom sky with torrents of rain. The earth quaked hard leaving a deep dark gash sucking in water, flotsam, and sand. All life itself. Drained. They felt the ungodly shake all the way to Honduras. The water tower collapsed. Buildings began to crack and crumble. More things swallowed up in the darkness. And the earth let out an awful groan through sheets of panic, sting, and rain.

All that next day. A cheerless wind. Whining and whistling. Loud enough to break your ear drums. And a harder driving rain. Relentless. Then Wallaceville disappeared. Impossible. But gone. By raging surge. Or it sunk into the ground. Sucked out. Sucked in. Sucked under. Around Grace, nothing left. Grace recited from *Di Buk*. Arms aching for holding on. Holding, curled she be, limb to purple limb and full of sandy snots and grime.

And that's when Grace's house blew off its stilts. It floated on the crest of a great surge from seaside to lagoon. Grace sick with the dark sounds of the shrieking storm. Felt it groaning through the floating house, smashing windows, tearing shutters, things falling, layers of *marley* (linoleum) popping. That whistling sound. And the smell. Shit. Gas. Death. Floating in it all. Grace feels that smell creeping into her and all around. *It renk*, she say, *Xtabay, di Devl servant, di Beast. Mek sens*, she say. *Xtabay*, the green snake turned beautiful seductress of men, with the skinny tiny legs of a turkey (Hagerty and Parham 2000: 19–25). Everywhere it goes it leaves that horrible stink and nothing but trouble. *Xtabay* makes pacts for men's souls and can make them rich. But watch out. *Xtabay [always] wan pay some weh or ada*, Grace stresses. It is the way a few local men got rich with some tourists, by making a pact with *Xtabay*, they say. Rich men, but the Flood, *Xtabay*, the Beast, a trembling assemblage indexing *somet'ing comin. Somet'ing trouble. Always somet'ing comin, gwehn to kalek*, Grace say. Trouble mixed with this strange storm set deeply into Grace's bones, her skin, measuring things frantic like a weathervane flapping in a hurricane. *Xtabay gwehn to kalek*, Grace laugh bitterly.

And the house landed in the lagoon. *Dehdeh in di mangro*, she say. *A football field away*, they say. *Impossible*, everyone say. Grace in that house for three more days nourished only by the Bible. Tilting on tides. Up and down. Stench hot. But she be cold and shake. Three days in cold Hell until they find her. Bits of her skin worn raw and festered. Shaking on weak swollen legs. Feet the size of balloons. Face cuts green with puss. Salt crusty mouth stink. *Jesus*. Bile, scrapes, and scabs. *Christ*. Things can't hold.

She, in trauma, tears with the telling:

Dat stink, shit and cloy. Stink, dat house. Doti clothes and hair. Greasy heat wet and oil. Me holdin'. Holdin' mi pets. Holdin' mi Baibl. Holdin' mi face. Mi house just holdin' to di mangros. Mi, holdin' to mi bed wen all watery hot dotinis: sand, serpents, and geckos. Somet'ing unholy. Somet'ing in mi ear wigglin'. Holdin' hungry, she say, in tiny tears.

Bumpy crack up, she say.

Watch mi house float oava di bering grong. See mi Daddy's kaafin com'n up [along] side. See he silly face. Grinnin'. No teeth but lat a bite, he insinuates himself once again into her life and she shudders hard.

Her daddy's body was found on 10 August 2000, out on a caye, eaten up and rotting under a Ceiba tree. How he got there and when no one knows. He hadn't been seen in the village at all for months prior but some say that he was hanging around with some strange tourist woman talking crazy and acting crazier. Then he disappeared. *Xtabay*, some said. Seduced by the demon-beast who took him. Xtabay might look pretty but don't be charmed by its striking beauty or with the bargain it wants to make with you. The malicious creature preys on Belizean men when they are sure of new money or new success. Xtabay drives them crazy while it lures them back to the Ceiba tree home where they always meet a grisly end. They entombed Grace's daddy's remains in the village graveyard, in a box, on top of a piece of ground that always shook, until it finally shook loose and open during the Flood.

Mi Daddy hissss like di beast, "You gwehn to die gyal, yesssss!" He grinnin' big, now she shake with this dead man's provocation.

I see Miss June house. Float. Turn. Upside down. Fallin' in. Sink, she weep while stomach turn.

Nuh houses left to see. All swallowed. Sunk. Pieces floatin' an di sea. Gyaabij. All chrash, she sob. Bile build.

Mi hold fi dear life and pray," she spit cry, things collapsing crazy.

House. Floating. The Serpent-Beast-Daddy-Xtabay. The surge. A Flood. Devastation. Collapse assemblage. This, the other side of Paradise in Belize that Miss Grace still lives with and that I take to be an activating line of flight, something immanent of an odd movement. And when such movements are "captured" in processes of reterritorialization they constitute new territories conjuring a creation, a something otherwise (Deleuze and Guattari 1986a: 130–31, 135–36). For Grace, Paradise is God's pure act of creation, *vita*, life's perfection now becoming lost with the tourists. A crime, *vitum*, by human action and greedy hand, tourist money, the vice that carries with it destruction. And so again the Flood, *vita-vitum*, the impossibility of living creation yet still a happening that ingathers multiple beings in entanglements. And so, on again, but with a difference. The point about Graces's *vita-vitum* assemblage, generated out of her Bible teachings from Revelation, is that it makes possible a process of thinking otherwise of the processes of Paradise productions as stable identities, territories, and structures. Those are secondary formations upon which the primary processes of "becoming-other" creates something durative out of bodies that are, for the moment, directly touched by the forces of chaos as a plethora of orders, plateaus, strata, sensations becoming Paradise, something extracted, becoming expressive, a provisional form, enough that it might sustain Wallaceville as a nervous intensification and a possible, make-believe space.

My focus is on those primary formations of difference, repetition, and intensification that constitute the forces of "becoming-other" in Wallaceville and in Belize more generally. Forces of material indeterminacy cohere with the forces of living, not just human, bodies to exert the production of *somet'ing*. Not so much part of a system of signs or a habit, for Miss Grace, this *somet'ing* acts as an intensity organizing into a refrain toward a worlding, a movement of things that enables creation.

And no one died in the Flood, except for those nineteen tourists on the tourist boat who partied to their deaths. And that said it all. With that. Grace shaking. It was said. Grace and her Bible saying it. Constant companions. *Di prais dehn pay. Me. Saved by mi Baibl, fi chroo*, Grace say. *Di tourists find di Beast*, she say. But Grace, attached to cat-dog-parrot-Bible. All of that. When nothing in Wallaceville was left at all. Just sea trash and death, ripped up houses, smashed sunken boats, debris of all kind settling in dirty water where houses once stood sturdy. All stink-rubble now.

Grace's close friend Miss Gloria said that nothing in Wallaceville was ever the same after the storm and earthquake, what everyone simply called *The Flood*. Along the beach, Grace and Gloria remember that as children they used to gather together with other villagers on those hot breezy nights sleeping under thousands of coconut trees and a canopy of stars. Beach fish fry and family fires. Visitors were welcomed but there were only a few places for them to stay. So families started to build attractive, colorful cabanas.

Most say that things started to *change up fast* in Wallaceville in the 1990s when Belize started to formally attract substantial numbers of tourists, in that time just before the Flood. That's when all sorts of new people appeared in the village and villagers began to get caught up in chance tourism-business encounters that seemed mostly to turn sour. In all of that, these encounters activated some new gesture of experiment in a local-to-stranger feel and movement crucial to making those *crazy crazy* moments Grace feels. Nevertheless, the village grew with tourist resorts, new money, good times, deception, new tour operations, less family cooperation, theft, drugs, nudity, lots of drinking, lots of "new."

But then the Flood. The post office. Gone. The social security office. Gone. The school. Gone. The church. Gone. The village council. Gone. The tourist cabanas and local hotel. Gone. Roads. Gone. Almost every house. Gone. Power. Gone. Trees. Gone. Beach. Gone. No infrastructure. No architecture. No social structure. No food. No Paradise. All gone. All of life, dreams and memory, spread out in bits and pieces across miles of rubble. Bodies in shock as waters receded.

Out of Paradise, all hell-forming now. And no return. Life and conditions permanently altered after this. But return, villagers did. Everyone picking through rubble for days. With the signs of the Beast now upon them taking the sweet breath out of everyone and everything. And things all in gasps for clean air and a clean start. But no. Grace read:

> And the heaven departed as a scroll when it is rolled together, and every mountain and island were moved out of their places. (Rev. 6:14)

And Grace knew that was just the half of it. It was a sign. *Gaad almighty. Mek di sign dear Laad?* Grace can't shake her Daddy's awful laugh. The green serpent's slither-loud laughter, lousy with awful stench and thicky thick air hisssssss. *Xtabay*, Grace say. The green snake. This is how it goes. Flood. To Serpent to beast Xtabay. Nature transforming, this time within the orbit of a cursed village, Wallaceville's sinful greed and lust.

Grace's Daddy had an affair with a Maya girl. Grace say they would meet in a hiding place along the lagoon under the big Ceiba tree where people fish in the evenings. Later, that green snake passed over the lovemaking spot and sucked up their moist sin. That's when the green snake can take on the Devil's power to turn into Xtabay, Grace shudder. Where you find Xtabay you find new seductions of surprising gifts, but it is all troubling riches, complicated cash mixed with a surge of ecstasy and desire. All bad. All sin. During the Flood. When she thought she saw her Daddy's rotting coffin coming up, Grace got *some strange, crazy feel. Daddy kip comin' kloas. Turn mi blood cold fu he smile like Xtabay's servant. Feel di Beast.* "Our Father, Who art in Heaven." Me. Frozen. *Lose mi breath but di prayaa work.*

That's why Grace made it. She prayed with the approach of growing fear: Flood, Serpent-Devil-Beast-Daddy-Xtabay. Bad weather, tourists, new riches. What she conjures in the waves of intensity that flow through her shivering body, what she cannot completely name, but what assembles in wild feelings of sadness mixed with elation, confusion, laughter, and furious fidgets are all intensities of Xtabay-beast-time change that grow tactile. And she asks herself what does her prayer touch? What does her impossible miracle of survival conjure? The praying itself, an active work of deferral, like interest paid to the Creator-Spirit on an acquired debt that drives right into more intense nervous exchanges.

Miss Grace say to Miss Dot say to Miss Thurnicia say to Miss Runcia, that it is the Beast she fears the most. It's what *fears* them all. They get the *fears*. That feeling fills them with stink sick and dizzy panic. *It hurts da heart, crawls ovah skin*, Grace chokes. The Flood. A warning. Don't go there. For Christ's sake turn around. *Bak to Paradise. Way bak den? Too late? Too late den. Too late now*, Grace laments.

Grace says that this is just the beginning. Clean up what is left. Clean slate and try again? But the Beast is upon them. And who shall stand on that golden shore? Who? She says the Flood was only a sign. *Flood was nothin'*, Grace say. *Lamb hear my prayer*, Her guts turn and she gasps for air. Her Obeah curse ruptures air. Weeping. Flesh eyes wet before they dry not. No innocence. Not for anybody. Not for anything. Not now. What else? Not now.

And Grace can still feel those sticky grit-sick connections today. Grace's immersive encounter with the Beast as something unfolding, something generative that she felt in that floating kinetic moment crossing through and over things in her house, something let loose, water waves touching waves of shivers, touching waves of intensity, touching waves of change. In movement the consequences are not clear. It's the *what's next*, or the *and then* that are emergent, rogue and *crazy crazy* that are her attunements. And that's what *reach* for Grace.

> And the serpent cast out of his mouth water as a flood after the woman, that he might cause her to be carried away of the flood.
> And the earth helped the woman, and the earth opened her mouth, and swallowed up the flood which the dragon cast out of his mouth. (Rev. 12:15-16)

You Betta Belize It: Beast Signs

The tourists started *coming fast* in the 1990s before the Flood, steady for ten years maybe more, changing everything in Wallaceville. That's the way locals put it. Everyone was on board to take advantage of new government policy that intensified around a realization that the country had little else legitimate but its natural environment and cultural life to attract capital and sustain economic life. So new campaigns were instituted through the Belize Tourism Board to attract tourist industry capital. Finding tourist capital is like finding a pot of golden coins. Stranger miracle money? Xtabay? But for those like Grace who are not seduced by the tourists and the industry, the tourists are a big warning sign. When carefree, middle-aged white women started stumbling nude onto the beach. That's when their wealthy husbands and boyfriends started thinking seriously about tourism development, dive boat operations, developing caye resorts, buying big parcels of land for high-end golf courses around new international airports, casinos, and resorts while moving into the "simple life" of tropical Paradise with downtown demands for comfort and happiness. Somebody's money was *flowin' in*, Grace says. But whose? And how? What else?

That's when local men gave up fishing, eventually destroying the most successful fishing co-op in Belize, to take the new, official, national tour guide courses and exams to "license up" and become tour guide specialists, a few boasting the charms of local life with cunning smiles, others advertising "knowledgeable services and guiding," still others to become Rasta beach boys selling ganja and taking chances on performing the "local native" while taking up all of the gendered and primitivist tropes of a tropical Paradise they could make use of. Making lots of new Paradise money by doing it and living on and off better than ever and *drinkin' Belikins* to some laidback Reggae tune fast becoming the generic "beach noise" soundtrack of Caribbean excitement and attraction. *Here I am, fucked, but me hapi*, Stretch say. It's never all or nothing, for or against, but a continuous flow of tourists and paradoxes, detours without full plan or purpose, uncanny facts, slippery like contra-histories.

Anxious, Grace could feel nothing but trouble ahead. Beast signs in new pots of tourist gold, shimmerings of dread, fear, confusion, and mystery, building, intensifying, and bringing things to life through whatever emergent image was selling at the time. And in whatever capacity, some aliveness was shaping up, some becoming monster was forming through the touch of tourism gold that was drawing new relations, seductions, and encounters as remarkable happenings.

Beast-signs. And the trouble was right there waiting. Some people like Grace said that the Flood is what you get for taking those pots of tourist coin. Not so much retribution yet, but a stern warning. It changed everything. She feel it. Crisis, like an intensely nervous *somet'ing* that felt like full-on trauma. *People movin' like zombies*, Grace said. But this was Wallaceville trauma, lived from oblique angles that moved into focus as much from the blunt forces of the Flood as it did from Xtabay's historical forces as a compelling example that conjured feelings of dutiful dread, the poignancy of everyday moral uncertainty, and the proliferation of feelings of concern and care against shock and chilling horror and resignation. Resignation and hope somehow mixing and swerving to produce feelings both unpredictable and difficult: trauma otherwise.

Without any money who could rebuild after the Flood? Over the next long while confused village bricoleurs played tour guides to make enough cash to help repurpose salvaged debris and to keep busy building shacks where they thought their houses used to stand. It wasn't enough. And precious little help from Belize City. More from the British Army. And Cuba.

Supplies, medicine, doctors, and labor. But little else was offered and few could afford to offer more. And those who could were perilously over extended. Everyone numb. Something uneasy set in like a grip tight around the throat. *Can't catch a breath*, Grace complained. *Slow death*, Lauren Berlant (2007) replied.

Wealthy tourist entrepreneurs managed to build or rebuild quickly. Beast connections, Grace was sure. As if they had some magic pots of gold coin to bankroll their efforts. Before you knew it their bars, restaurants, resorts, and hotels were "up and running." They had their new houses under construction just as quickly, and much nicer than before. The banks reopened and bank men made offers to locals that few refused. That's when the well-dressed smiling storm trooper land developers moved in with bags of money. Money that felt to Grace like the Xtabay's buried treasure. Pots of newfound gold coin encounters. Realtors reaching for legal contracts out of the back of their big red Land Rovers. That's when an army of pushy surveyors measured up named and titled land and bought it from locals on the cheap, taking advantage of needy families who took those pots of gold.

Eagerly and reluctantly at the same time new magic money attached to a baleful future present just beyond reach, working in the blink of a moment, out of sight but in the visible shock waves of village trauma. That money was to finance the new tourism schemes that eventually, with a few notable exceptions, pretty much failed to grow and finally died. That helped put the strangle hold on the future. *Nuh breath left. Signs a di Beast*, Grace stuttered again. That's when Belize loosened its banking rules to net the flow of international capital that started to float in like newfound *treasure* throughout the Caribbean. Greed was everywhere and into everything, on skin, breath, smell, and voice, impossible to place but deeply felt as something settling in all the same, something that urgently needed a name that wasn't the Beast's and did more than materialize like a forceful sound that nervously hung on the tip of the tongue struggling to find breath and form as words, a language, an explanation.

And that's the time when Belize deregulated and radically transformed its telecommunications infrastructure to make banking and retirement and land investment easier. That's when the international invitations that came in glossy images of Paradise went out to the world seducing those retirement lifestyle transients with money to buy up the simple pleasures of tropical happiness. That's when locals became "natives" (see Holmes 2010) and Belizean nature became Nature, and Belizean postcard images became invitations to a friendly tropical good-life Paradise. Ubiquitous fixed happy looks of the locals that passed from friendship, to resentment, to disgust, to rage, to despair. *You 'tink we betta dat way? Playin' di Beast game? Suppose to be all service now, not servitude?* Garvy spits. But there was lots of exciting action riding on these seductive invitations to Paradise, when first-time tourists returned as delirious "lifestyle migrants." And all of this resting precariously enough on a fragile infrastructure and beach ecology hardly prepared for any of it; "it" being a cleaned-up, post-Flood, ecological dreamworld disaster area that would remain in perpetual and struggling recovery mode: a losing battle. Natives at a distance become the service sector for your better dreamworld in this fabulous Paradise.

The post-Flood official Belize government surveyors took all of their measurements from the original colonial Wallaceville survey stone markers (commonly known as the *Royal Stones*) they found still intact, in the ground. They redrew property lines of title from those three remaining reference points. No one, not even those who stood to gain, really ever thought that the new lines felt right. After that all surveys were immediately suspect. But their numbers counted. Impossible to measure memory, tensions swelled out of the impulses of bodies moving across the land, "flirting with space" as they move with the changing texture of sand on the feet, how from your hammock you could hear Auntie's voice from behind slatted windows, smell her frying fish in coconut oil on the right breeze (Crouch 2010). Nevertheless, many sold subdivided or sub-subdivided titled land using those new surveys and the new "paper" that guaranteed them. But everyone still disputes the new property lines, turning brother against sister, relation against relation, friend against friend, previous social lives and responsibilities against future individual opportunities. And the new landowners, tourists with the money, make their Paradise dreams out of alarming land deals.

A complicated uncertainty of disaster trauma is the emotional terrain over which this story of remeasurement practices takes place, even today. This is life in the impasse (Berlant 2011; c.f. Povinelli 2012), or how life in Wallaceville ricochets wildly between truth and advantage, threat against aim, feeling like something is afoot but not able to force prediction into being, sensation into signification, and so life forms as a pestered temporality or historical capaciousness, an intuited sense of becoming-present.

Belize It or Not: Paradise

But not there, not yet. This is how beast-time is at hand and instantiates itself in Wallaceville with its sense of a sped up temporality, smoothing things over or cracking things up again, but coherent enough anyway to seduce calculations of local and tourist industry efforts in order to make more sense and cents of them; a matter of a continuous revaluing scene and experience against what to do with an opportunity. Caribbean capital time speaks: *You want Paradise pleasure? I'll give you all the pleasure Paradise that money can buy.*

Just keep your eye on the Beast. Get busy making beast-time value. Enjoy those Wallaceville transformations into an international tourist seduction. That's where the money is, where the future is, and with it, all of your enjoyment and peace.

If there is a generative weave of things felt and done in and as Paradise, it trespasses those clear distinctions between subject and object, something that unravels its subject, as a nonunitary entity, and something that exceeds its object, generating beast-time, a Belizean beastly "untimeliness" (Pandian 2012). Untimeliness, present and virtual, is when things and beings come to differ from themselves and in the space of that difference arises, contingently, enigmatically yet persistently, beast-time's durative quality. It is about how things continue to exist as potential in any Paradise actualization. And in "the emergent yield of its duration . . . life develops into unforeseen forms . . . through which things affect, modulate, and transform themselves and each other" (Pandian 2012: 549).

This is the Beast's unexpected time-signature, its durative, a "rhythm without the regularity, or a readiness to arrive" (Massumi 2002a: 20). Beast-time as an emergent vitality of felt suspicions, hopes, loves, and hates that gather force over time to conjure generative spaces of an unfolding, the vitality of things and beings in their singular acts of creative expression, a kind of "mattering" for better or for worse as an extension and an intensification or, sometimes, its suspension. This is the way the beast-time milieu is generated. This is the time of emergence, of excess, tangents, and "lines of flight." It ruptures linearity in a flood of sensation and sudden conjunctions that is the stuff of liveliness and potentiality, of creative and passional flux. This time of an otherwise endures in the face of crisis as metamorphosis to multiply the signs of existence itself without a telos but as a force-fullness in the life of a contingency.

Time-flux, like the feel of anxious betrayal when someone sells out or stops sharing news or new wealth. You are in it for yourself. Find your own tourists, your own salvation. It's entirely up to you to take advantage of things. But such moves and bodily flux conjures sensations (tinglings, headaches, body pains, a case of the nerves, instant waves of rage) on the way to a feeling, in things and beings that have to be literally tracked through dense and disparate scenes of desire, dissipation, seduction, law and labor, or in everything of which this new Paradise is fashioned. Such flux can be sensed in that feel of despair mixed with suspicion mixed with the hope that whatever tourism industry interest was buying up the land for development there lodged a promise of a good job, a better capital opportunity. Everyone scans the place for signs of some impending *change up*, some odd logistical moment, a new "for sale" or "help wanted" sign, chaos in the Village Council, or the possibilities and conditions that come with a job offer, even to rake the beach.

There is an apocryphal story Grace shared with me about a prominent villager a few years back who was said to have found a pot of Spanish gold coins in an ancient Maya site. He called it pirate treasure, and he found it while beached for a night, digging around on one of the little local cayes. Which one, he can't say now. He's dead. The Flood turned up the surface of some of the cayes in such a way that they revealed such long-kept secret treasures and he found one with little effort. He exchanged the gold for "Mexico money" and then changed it back into Belize dollars to shake off traces of Obeah. Not stupid at all. Secrets. Power exchanges. Subterfuge. Anxious struggles over who is going to win and who is going to lose. *He talkin' too much, braggin' 'bout new boat and motor, buildin' a big cement house, nice truck, but da man have no business sense or no way fi makin' real money*, Grace say. Almost overnight he was rich, and many families speculated that he had found one of those pots of gold, Grace recalled.

And then there was no *make nice* with neighbors, friends, with anyone. He could grow instantly angry in a flash with his family, but especially with friends and relatives who would ask him for loans. He grew solitary. He got drunk a lot and found a bunch of new tourist friends to hang with. He got into fights. Then more fights. Bad fights. Then there was that fight to the finish, down in Mango Creek, with knives and a baseball bat. That killed him. Xtabay treasure, the twist and pull of confusion, chaos, greed, dead relationships, sour sentiments, sad sensations, another dynamic in a village feel of things fashioning the beast-time. And no one knows where the rest of his treasure-money is hidden. It wasn't in his bank account, his wife said. And so everyone is still looking for it, senses alive for any sign of its presence, an Xtabay event maybe, scanning for what else or whatever might become present, felt, eventful. Paradise.

In Wallaceville right now there are a "precious few pieces" of titled family land left. Titled land is valuable. But just ask to buy it. What's left of it is all not so secretly for sale for an undisclosed number of millions, even when many of them are hotly contested properties hooked to family disputes that may never be resolved, even among the Anglo-Creole elites. Owners sit on that land hoping for some big hotel chain, a church, some expat property investment group attached to a Belize elite family fortune, an NGO, or some rich tourist to offer them some miracle deal, the impossible Beast deal a local won't be able to resist. But it takes real Beast power for some business interest that is big enough to settle for themselves any legal or political encumbrance, pressure, or problem. And even the dream of moving onto the family caye, the last refuge, and out of the wild grasp of the *Beast race* can turn on you if the price is right. But then what, once all you have is the money? And that stops everyone cold in their tracks. Selling guaranteed owned property, land with paper, is the last growing pot of gold coins, the last source of "big bucks," left for those born in Wallaceville, as resort developers and lifestyle retirees make deals for it. *Aai ya yai bwai. Dehn sellin' fi dehn souls to di Beast,* Grace slowly and sadly gulps, *One way or ada, fi chroo.*

Pride turned to panic morphing into pain, doubt, shock, confusion: a beast-time feeling of a Paradise structure in the making. Trust no one. The courts are plugged with such uneasy disputes that linger on for years until the cases burn out or the banks pull what they can out of a bad situation and put the property up for auction. The weekly newspapers are always filled with auction announcements that locals scan closely for Beast-sign deals. Or disputed properties are left to rot into ruin as transnational financial backers get nervous and pull their money, or the market tanks, abandoning big condo projects and leaving them to be ransacked by those looking for free window frames and siding and sliding doors, and copper wire while those tourist investors who paid into the projects are left "high and dry." It's all in the timing, in the international economy, in who you know and who you don't know, yet.

Everyone can feel it, when things go *crazy crazy* and spin out of control. You can feel it in the new atmospheric pressure building out of signs of speed, greed, population intensity, new stuff, stories, and secrets. All of a sudden, and unannounced, ranks of jumpy survey teams show up somewhere on the new sand beach to measure and remeasure a piece of land whether it has a house on it or not and then leave without a word, the land left dotted with survey flags and red markers. Each team has its mission based on a worried client's (almost always a relative) motives. It's mostly then a matter of someone's memory, measurements, and legal documents against someone else's bad life decisions, public secrets, crazy new forms of life entangled with creepy new social relationships and marriages or just plain crazy new family interests. And no one is in the mood to *make nice* any longer.

The result is beast-time land multiply measured and stratified legally, conceptually, emotionally, and socially. Sharp-focused new international interests rubbing up against ragged memories and what was once another sense of place, now romantically solidified in an image of pre-Flood life and times, to become the newly created spaces of hot deals, public secrets, crazy dreams, ludicrous schemes and cold disappointments or Paradise tropicalizations. All this suspicion. Tension. Intensity. Craziness. All of it constitutive of post-Flood life in Wallaceville. And all you hear today is loud construction noise as one tourism resort project after another pops up. This is the newest beachfront soundtrack in Wallaceville. The newest sounds of the Beast growl are added to the thumping beach bar noises and the raucous restless party voices of the tourists getting high while generating Paradise mixed in with the growing army of suspicious migrant construction forces from Honduras, Guatemala, El Salvador who are the disgruntled labor building it.

This new tourist industry restructuring of life has replaced the failing fishing economy. The mismanaged reefs are out of *just 'bout every'ting livin'*, Grace says, leaving the place in the hands of a new economy enlivened as an ecotourist dreamworld of tropical *take nothing but pictures, leave nothing but footprints* Paradise. The prime directive? Get with the picture. Attach to the affects of a tropical tourist pleasure spectacle. Work those ecotourist tropicalities into their sedimented, territorialized tropicalization plans and into their incentivized eco-management schemes aimed to maximize eco-system-tourism service maxims: Stay off the coral. Kill off the lion fish. Save the lobster. Save the conk. Protest the cruise ships. Visit the ruins. Enjoy the arts festivals. Respect the *kolcha*. Buy local and let them make the profit. All the while Wallaceville gossip grows intense with hushed street whisper rumors of locals making secret private land deals with big tourism developers or with members of the political elites, or both. Local eyes and ears are everywhere and always alert to plots, deals, magical pots of gold coin and tense magical Obeah connections that cross the nation through Twitter accounts, Facebook pages, the banks, and personal cell phone text messages revealing, or not, next moves, secret associations, strange connections, linkages, and networks of the craziest sort, heating up public discourse, raising personal suspicions and family misgivings.

Life in di hands of di Beast, Grace trembled, watching some developer start up a massive condo development just across the sidewalk from the house her family finally pulled out of the mangroves to repair and restore on new posts after the Flood. Priceless beachfront land carelessly sold somehow after Crazy Mojo was shot to death. In a strange move that seemed an impossible one to make, two lost Belizean relatives from "who knows where" with deep pockets, some local links, lots of strange political influence, and secret new international connections somehow bought the land without Mojo's knowledge. Stunned, Crazy Mojo had to move. He wouldn't. The police were sent in. Something happened.

Some said that Crazy Mojo was shot in the back as he stood in protest to being cheated out of his home and that his lifeless, naked body was dragged away in the back of a police truck. Some said that Crazy Mojo, a smart, decorated US marine, had a lot to gain, really, and that he was working all along disguised as a mad man with some elite intelligence-political group made up of faraway relatives who stood to "make a killing" on that land and that he simply disappeared and now lives like a rich man in Miami somewhere. The stories of Crazy Mojo, the naked man taking outside baths in his little red plastic wash tub, the guy who used one of the two rooms he lived in as his toilet, working incognito as a spy for tourist industry cruise ship land developers conjures some of the absurdist qualities of beast-time oddness as impossible everyday moments through which the Wallaceville ordinary emerges. *Mi tink we all goh crazy*, Grace says. She says she is watching the beast-time unfold right now from her porch and that she now has some idea of how the Beast operates. She can smell the beast in the drifting smoke and burned cinders of Crazy Mojo's house, an agitation that drags things into view for her as a momentary excitation that emerges out of something that smells like burning garbage on the beach, of things becoming resonant in the tightening grip governing everyday sensibilities.

Then there was the gossip that no one believed until they had to. About what happened to old Mr. G. after the work and money he put into rebuilding his hotel after the Flood when everything he cherished either floated away or sunk into the ground. He made it a great success, his pride and joy. Maybe too much joyful boast, some say. Mr. G is known for his *loud mouth, like it a disease*, says Miss Grace. Well, this old guy from Belize City shows up with an old piece of official paper notarized and legal showing Mr. G that the land he rebuilt on, the land his father had given to him on his return from the States in the 1980s, was not his at all. It had been sold years ago when Mr. G's father traded it for a precious and necessary fishing boat and motor he used over the years but that everyone forgot about. Mr. G was legally forced to vacate the land leaving his prized hotel behind while the old man from Belize City and his son brought in a bulldozer to flatten the building and the dream. That was the end of that. Except now the land is for sale for millions, leaving doubly traumatized locals like Mr. G in complete despair and destitution.

Meanwhile up the beach a famous Hollywood director built a high-end all-inclusive resort to go with those other monster properties owned by international investors interested in pioneering experiments in gated-community, dreamworld-Paradise living. With a helicopter pad and a huge illegal pier for yachts, the Hollywood director attracted the high-end tourists and that attracted those who were attracted to the high-end tourists. All of a sudden there were two new high-end subdivisions, and then others, and still others. And precipitously Wallaceville was on the international tropical eco-tourism map. More money appeared earmarked for crazy timeshare housing dreams, sunrise-beach- or sunset-lagoon-side. Old Flood-ruined resorts were converted into fancy new ones with the latest state of the art infinity pools, spas, restaurants, and bars. The rest is slowly, sadly rotting into beach rubble, all serenity gone. Wallaceville is back on track today to become one of the most attractive "go to" destinations for sun, sea, sand, and beachside blue sky relaxation, *for everyone around the nation and beyond*, as Love FM advertises. Visit. Retreat. Revive. Relax. Retire. Come and go. Come and stay. Do as you please. Echoing that laid back *don't worry, be happy* ethos.

But there is more to power in the Caribbean postcolony than that. Belize is located in the turbulent, wild, and messy Caribbean currents of global empire and whatever that means in the making of life there and whatever characteristics analysts routinely use to define social, political, and economic life there, they still can't begin to describe the unfolding generative moments of an emergent life, when the senses become jumpy and everyone alike gets a case of *the jittas*, or they are *taken* by the sublime hilarity of crazy encounter moments, or they are suddenly braced by shock, or impacts, or they feel secured by the promise of an attachment to things and the "cruel optimism" of interests, or by the sudden sad disappointment of one cherished thing or another turning unsteady, fugitive, seductive, sad, or violent (after Berlant 2011).

Active forces and reactive circuits of forces build up some sense of a commitment to one thing and then, in confusion, allure, seductive attachment, fear, charm or desperation, to something else and in the movement *somet'ing* tactile materializes. Or it's the way people just vanish because they want to or need to, out of desperation, protection, embarrassment, or because they are seduced by some new life plan or attraction. Such is the unpredictable liveliness of life in the grip of the Beast, acts of Wallaceville realization, and in these actualizations, things unsteadily drift into other scenes, refrains, and fields rubbing up against yet other practices and go eccentric or excessive or otherwise. It is to the otherwise that I turn my attention. The otherwise animates things still uncaptured, things incommensurate with their capture as an ideology, a sign system or a system of representation of life in Wallaceville. Ordinary life is much more fractious than that, too unsteady, caught as it is by the force of its impacts, its intensities (see Stewart 2007).

Belizability: Impossible Expressions of Paradise

To think is always to follow the witch's flight.
—Gilles Deleuze, *What Is Philosophy*

To this day Grace cannot shake that impossible moment during the Flood when her house was set afloat over the village graveyard and into the lagoon. She felt a special vulnerability, threat, shock, and the strongest sense that things would never ever be the same again, that things had become so eerie and frayed, unnerving and disturbing as to form a sea change and a hard fall with little sense of reprieve every so often in a wry sense of the hilariously absurd. *Beast-time*, Grace calls it. And she feels the Beast still. Grace will never shake that dreadful experience during the Flood when she saw her Daddy's open casket floating by. Him smiling his stink-horrid toothless grin. She felt it happening, insists that it happened. Everyone else, who knows Grace to be a smart, no-nonsense elder of Wallaceville, tightens up and finds *nuh comfort* when Grace starts talking about the experience. Her friends and children worry about her. *She feel di mess-up*, Gloria says. But Grace can't shake the intensity of her dead Daddy's smile or the horrible hiss of his wheezy screech-voice telling her, *You goin' to die Gyael*. Here, in this scene of haunting attachment, Grace puts things together as a composed tactility touched by a perplexing metamorphosis—her dead Daddy to green snake to Xtabay— inciting the beast-time, the conditional present, the future anterior of an emergent Paradise. The condition of time through which she now moves is in visceral complicity with this New Beginning, rocked by her story and redistributed by a tidal flow of its tellings and retellings to some differential rhythm that repeats itself without rule that still moves things madly along.

And it doesn't help that of all the coffins and bodies that were wrenched from their resting places in the village graveyard during the Flood and were found and put back in place, that it is her Daddy's body that remains lost, and so its absence is still a mysterious, gruesome force, an unnerving vitality full of possible impacts. Fragments of the rotted pine box that held his body were found here and there floating in the lagoon, but his body simply vanished. They say the lagoon crocs must have finished what was left of him. But Grace still feels his awful presence. She feels it when crossing the lagoon and things grow suddenly still and cold. On occasion she catches a glimpse of him from between the slats of her kitchen window; standing in the yard. The first time it was just after her house was put back together on its new posts. It was a very hot night and Grace couldn't sleep. Out of bed for a glass of water, she stood at her window looking out at her mango tree. The sea was quiet. There was no breeze. Still silence. One of those nights when, with every footstep in the sand everyone pretty much knows whose steps they are by the heft and the grit-grind sounds and pace of the steps, by the breath, prayer, song or by a hushed *g'night, g'night*. Yet undetected there he was, standing under Grace's mango tree staring up at her, wearing that gruesome smile. He hissed a stink sound. But so cold quiet. *Shhhhhhhhhhh*. She slammed the slats shut. When she opened them again he was gone.

And now, others are haunted by him too. Grace says one of her grandsons, ten-year-old Danny, feels him, and with that haunting sensation another crazy moment formed. On a warm evening just a while ago, when the sun and moon got lost in each other at the very verge of darkness, while fishing his favorite spot on the lagoon shore, close to that same old Ceiba tree where Grace's daddy had his affair with the Maya girl, a strange old man passed by quietly and stopped to watch. No name, a few wiry strands of long hair on a pale green and bony skull, rotting mouth, no teeth, all stink, in rags. *Bwai?* he barked. His voice turned the spot still-cold. Danny froze. He hissssssed and circled. *Bwai, come. Mi have somet'ing fah yuh*, the old man spit. Danny looked down at his hideously skinny legs and tiny feet. *Money fi yuh, bwai*, he sputtered. *Xtabay servant*, Danny whispered, *like Granny said*. He couldn't move, until he could. And then he scrambled away in a panic and into the arms of his grandmother, sobbing inconsolably, shaking madly. It gives Grace the "shakes" just to recall that telling moment, and the clammy cold she felt on poor Danny that she couldn't thaw with her embrace. It haunts them both still. It haunts everyone else who knows the story. The Beast, the Xtabay, Grace's Daddy demon servant, pulling things into a crazy relational alignment as a moment that matters, as a mattering. The Beast assemblage keeps Grace *jompi*, living life on every other side of ordinary, alert to things of the threshold between the living and the dead, unsettling things that will never make sense to her whether she likes it or not: haunting and a little bit out-of-control, tempo without meter, beast-time emergent. Others feel it too.

Grace's habitual fretting about her dead Daddy, her grandson's experience, family haunting, village wide precarity in the time of tourism, life conditions, complicated Obeah protections: this entanglement rubs her senses to the bone. She knows that she is into the beast-time now. They all are. With Grace's help I track how beast-time takes form, how it pulls some collection of forces, sensibilities, rhythms, materialities, and situations into contingent alignment to become nervously generative of *somet'ing*. How do such temporal-sensual textures of experience find a pulse in a haunting, unnerving ongoing and emergent tourist present called Paradise?

Chapter 2
Impossible Tropics

Money Grows on Trees?

There was this late April afternoon in Paradise, 2001 version, a few months before the Flood, that no one will ever forget. That's when Wallaceville was showered with US dollars. *Lord, the money was blown' in like manna from heaven*, Miss June exclaimed. This was another mysterious arrival, unrivaled in its impact as an arresting force, at once vastly familiar and seductive and entirely uncharted, even shocking. Familiar and seductive, because this has always been the promise of the state and the tourism entrepreneurs—that money would flow *like di air we breathe* Richie insisted—that with more tourism everyone would benefit; uncharted and shocking, because, among other things, no one expected money to flow quite in this way. More mystery and intrigue. More giddiness mixed with dread.

This is really the story about Mr. Richie who, while fishing one day is said to have found a bale of cocaine floating in the water off of South Water Caye. It was said to be tied to a stash of money said to be in the hundreds of thousands of dollars. It was said that Richie pulled it from the water and, once back on shore, hid the treasure up a tree. His nephew Bobby recounted the story, of which many locals now have versions even if they won't go out of the way to tell them:

> *Richie finds this serious fuckin' stuff. He hides it, like pirate treasure. Soon he looks like he won the lottery. He's not fishin' or workin' tourists. I know what he's doin. He's into the money and the coke. So life is a party. He starts hanging out with tourists all day, buying drinks all over the place. Relaxin' and talkin'; maybe too much. He buys a new boat and motor, top o' the line shit. He's lookin' good and wants to start a tourist dive shop business with me. Now everyone starts askin' how he can do that. Maybe the big drug dealers get suspicious. Richie doesn't care. He's gettin' high all the time. He's not payin' attention. So one afternoon it's really windy. Richie is drunk but he*

> *seriously needs some cash. So he heads for his secret bank in the tree, 'cause for Richie, money grows on trees. That's what he tellin' everybody. So that's got everybody lookin' in the trees for his money. People followin' him around all the time but no one finds his money tree. Richie, he's too smart. But he fucks up. He left the bag of money open or something and the strong wind . . . . All of a sudden, all this money, hundred-dollar bills, fifties, twenties, tens, it start blowin' all over the place, down the street, onto the beach, in the water, in the air, on the road. It's rainin' money and it's landin' everywhere. No one can believe it, but they're pickin' up bills. Miss June say it's like a miracle. Richie figures it's gotta be his money. Not happy. By the time he gets back to his bank in the tree, almost nothin' left, all blown away.*

It was just after Richie's money blew into town that locals like Bobby remember the first sighting of Mr. Pete. The unnerving appearance on the beach of a tourist named Mr. Pete, as if he were conjured out of thin air, was one more instance of life becoming impossible to live, to believe in, and to accept. Bobby guessed he was the drug dealer who must have heard about Richie's loot and was here to reclaim it and the coke. That's when Richie "got lost" for a couple of months leaving villagers and who knows who else wildly searching Wallaceville for any connections between otherwise disparate and unusual things, scanning for signs of wealth and euphoria, criminal threats, or suspicious behavior in an increasingly unsocial, uncertain, and chaotic world and in a life turned impossible that started to get on everyone's nerves. It is impossible to tell the drug dealers from the tourists, or either from the international resort/condo speculators whose side-deal scams were just starting to create the state of emergency called "tourism development" in Belize. Even then they were buying up titled land in Wallaceville for a song and dance and selling images of local culture as an escape to a pristine Paradise with laidback, friendly natives who were pretty much being *voted off the island*. Mr. Pete was a troubling local presence, another impulse-machine pumping a contact sensuosity into the Wallaceville nervous system, seducing and shocking locals and expats alike.

No one knows where Mr. Pete came from, only that he appeared from "wherever" to spend a few weeks compulsively raking the beach. And in early April of 2009 he reappeared to do the same thing. He always appeared on the beach, rake in hand, geared up with a Walkman and earbuds, orange swimsuit, and a big straw hat. Why he rakes Mr. Pete would never say, and that generated a distressful shiver of dread and panic that ran through the nervous system of a village that no longer seems able to keep up or on track with everything going on these days.

Mr. Pete's initial appearance coincided with Richie's money troubles and a new rash of violent physical attacks on resident expats and tourists, two new crack houses, and a variety of petty robberies. No one blamed Mr. Pete for the attacks or for the robberies, but his sudden reappearance and then disappearance and his odd behavior served to focus everyone's attention on tourist encounters in this "Paradise by the Carib Sea." These encounters bred an excessive exchange of stories about drug dealers, brutal local violence, other strange tourists, pirates, crazy expat locals, infidelities, theft, mysterious land deals, government corruption, tropical environmental collapse, and strange weather: flows and lines of narrative force that rubbed off on each other to produce a swirl of arresting tourist images of Paradise gone crazy, producing a friction that generated a menacing and jumpy energy of alarm that has seriously "roughed up" life in Wallaceville.

Tracking the affective intensities of life as it becomes impossible, a scene of immanent forces folding into a swirling assemblage of public feelings, means tracking the troubling state of suspense and suspension that haunts the place and its people and that lingers as a jumpy impulse trying to "make sense" of things that come into view as habit, shock, resonance, or impact. Lives in Wallaceville throw themselves together as event and as sensation, something inhabitable yet exhausting, a tropical dreamworld escape and an odd and haunting ordinary. Examining moments of encounter and lingering in the impacts of new signs of life becoming impossible means tracking sensations as emergent and potential emotional forces coming into play in this new state of emergency that is taking shape as neoliberal exception, on the edge of global empire, in Belize.

Impossibly Beautiful

There is a particular quality of light, sound, and touch as you approach the beach in Wallaceville. It lends itself to a feel for the place, the impact of which, as Jim and Annie said the first moment they laid eyes on it, "it takes your breath away." It's some combination of sun, sea, sand, and sky that intensifies an array of sensations as it dampens others. Much of this intensification and dampening is already encoded as advertising cliché, a carefully calculated indexing of pleasure in Paradise "to die for." In such cases these sensations stabilize momentarily into the commercial and cultural tropes of escape, Paradise, and natural beauty that Belize has progressively activated commercially over the past two decades. And yet the place still conjures a sensory impact that is almost more than a body can take or an image can contain. Your body builds its substance out of layers of sensory impact: drifting in the surf, skin impossibly wet, warm and salty, eyes trained on the light clouds passing over coconut-tree tops or on watching colorful tropical fish dart about the live corals, ears submerged in the gentle pulse of wave action that surrounds, buoys, and carries you along. Hear the underwater tingle *tic-tak-tic* sounds the fish make on the coral. Feel the atmospheric pressure of an ideal tropical day.

On the beach your body surges with the rush and flow of the waves, the push and pull of tides and sensations. Get with the picture, take a tour to the ruins, go diving, celebrate tropical life, maybe spend a day on the beach lying in a hammock, go to the gym, feel the muscle burn on the sunburn, and then do it all again. Your beach body knows itself by the strength of its liveliness, vigor, vivacity and by the seduction of its delinquent vitalities. The beach sand feels warm, powder soft, the water crystal clear, and the sky is a profoundly deep blue seduction. Helplessly, trying to take it all in, it's like your body can't totally collect itself. This deserves to be mentioned because it shapes so much of the immediate tourist sensations of this tropical place as "impossibly beautiful." Tourists are quick to pick up on these sensations and go with their flows. It's all more than something seen or revealed, and much more than a tropical representation can offer. Rather, it is something felt as almost overwhelming: it packs a punch. When you feel it, your skin dimples and tingles and your body is filled with an excitement you can't quite put into words, yet you sense it all the same in that breathtaking moment: a virtual-to-real move, an incorporeal folding into the corporeal, the body in movement, organizing as a tropical rhythm of life, coincident yet disjunct (Massumi 2002a: 11).

These feelings are never separate or exacting points of excitation. Each is fuzzy at the edges, open-ended intensifications, incipiencies, affects unlocking sensation's potentials as mischievous interactions, radiance spreading uncontained. The beauty and power of this Paradise dreamworld is felt in the play of sensations intensifying into something, just as feelings bleed into one another, and into a slow drift that builds these breathtaking moments into impacts: a body event as viral contagion, you pass it along with each excited touch, look, intake of breath, or smile. These are feelings in a state of emergence, feelings on the verge of their naming, still unfolding, fugitive, chaotic, and shifting. The body's give and take in relaxation, its impulses, its waves of giddy sensation and tension, are like the waves lapping the Wallaceville beach, each the same but different in their making and breaking, and in this movement, in the suffusion, there is a potential, a force gathering itself to a point of impact to instantiate something. It's a sensation scampering on the edges of a feeling, suggesting where a feeling might lead if it is left unrestrained.

This is the body as disjunctive encompassment, a kind of continuity but unlike one that follows a narrative line exhausting its signifying possibilities in meanings or as a type or illustration of some sociopolitical process. Here, the body becomes a continuous displacement of the subject, the object, and their general relation, creating and created through a folding and an unfolding of sensations freed of the terms that name them (Massumi 2002a: 51). On a beach in Belize, they are actual sensations, felt forces gathering to become something still impossible to describe although all the more felt. These tropical sensations are the forces of feeling that add to what one brochure helpfully calls "the impossible beauty of a Caribbean Paradise" (*Destination Belize* 2003). That's the tourist Wallaceville. It generates a "body without image," an additive movement from the incorporeal to the corporeal that registers as an included disjunction, or what Massumi calls incorporeal materialism (2002a: 60). It's like recognizing some feeling in your body that you have no name for yet or recognizing the feeling that grows in you when you have some name on the tip of your tongue but you can't quite get it out: wonderment and frustration, tension, movement, change.

A Case of Topical Nerves

I describe the impacts and the intensification of sensations in the space of a beachfront tourist Paradise but only to disrupt the vibe with the burdens of that same Paradise, conjured out of the unnerving reappearance on the Wallaceville beach of a tourist, Mr. Pete. Again it was Mr. Pete's second coming. A few locals with long memories remember when he first appeared in Wallaceville and began his daily raking ritual and then when he mysteriously disappeared, not to be seen again, leaving only a nervous tension in the air, an agitated buzz drifting through the village. That was in 2001, a few months before the Flood, and that was when Richie lost his money, and that's when a couple of big drug dealers seemed to make Wallaceville their new headquarters, and that was when rumors of the Rapture seemed to take a forceful grip on local sociality. Waves upon swirling waves of craziness. And then shortly after that the Raelians showed up. It was like one impossibly crazy thing overlapping with another but with no connecting logic, just a swirl of confusing and confused random crazy events that conjured a serious edginess that spun life out of control: life becoming impossible for almost everyone living in Wallaceville at the time.

Mr. Pete simply appeared on the beach one day and then a couple of weeks later he disappeared. Two times. What next? Mr. Pete, his name seems to mimic the sounds and rhythm of soft waves gently licking the shoreline against the relaxed, repetitive scratching of Mr. Pete's red plastic rake. Because that is what Mr. Pete did: he raked the beach daily with his big locally purchased red rake, through the morning and over long, hot, breezy afternoons. No one really knew where he came from in Wallaceville or where he was staying, only that he appeared from "who knows where" to spend his time compulsively raking the same few square meters of beachfront sand before he disappeared as quietly and mysteriously as he had appeared. Mr. Pete hardly said anything to anyone. He wasn't a talker or a socializer. And that, along with the public performance of his odd and empty ritual, once again generated a nervous intensification of dread mixed with humor that conjured a discordant shiver to run through the nervous system of a village. Wallaceville no longer seems able to keep up or on track with "everything" going on these days. Life feels like it has become excessive, too many things going on at once so that it feels all too much and it provokes a restlessness that irritates everyone and spreads as quickly as with Mr. Pete's raking antics.

Mr. Pete seemed benign, but no one could walk past him and his incessant almost manic raking without a comment or without taking the opportunity to watch his mute performance and to speculate, to look for signs of some crazy move or encounter. Everyone on the beach had something to say about him. It made Miss Gloria *fret*. *He mek wi crazy*, she muttered one day. *Wach (out) for he. Di man nuh right.* If Mr. Pete stopped his feverish raking at all it was to talk to the same three dispossessed locals. They seemed to be charmed by him and the charm seemed mutual. There were a couple of other strange expats whose everyday eccentric beach behavior (as ersatz pirates, drunken prophets, and nervous loafers) could be compared to Mr. Pete, but they fit the local description of "stupid tourists," incautious and rude but at least communicative. Not so with Mr. Pete.

A witness to all of this, Miss Gloria worries about her kids and her property and she is not alone. If only Mr. Pete would say something, give up his intentions, explain the crazy raking, hour after sweaty hour, day after day, but he proffers nothing. So villagers watch him and watch for him while tourists usually give him a wide berth. *Just another crazy tourist Ken*, Mr. Richie says to me one day. *Do you know him?* he chuckles. Richie's question comes with a tense laugh that belies his family's disturbing worries about everyday life these days with a village so full of strangers, strange projects, deals, and *odd stuff just happen'in* that they hardly recognize it anymore or know what to do.

Few could figure out where Mr. Pete slept or where he ate when he wasn't raking. Miss Gloria and several of her neighbors reported him to the police, but the police said that they had more important things to do than deal with a crazy tourist raking the beach. In fact Sergeant Ramos thought that he set a good example for local others. *They should follow suit*, he said. So most locals simply felt uneasy and puzzled by the performer and the performance. This meant that everyone kept an eye out for him, just in case. Someone said that he was this rich guy from California who lives in Honduras now, but likes to make side trips to Belize when he gets bored. But why Wallaceville? And that's when concerned locals tried to put two and two together as they started scanning the village for signs of bigger trouble: the drugs, the money, odd-looking strangers, strange looking land deals, the new stores, the big boats suddenly at anchor in the bay, the new cars, the kids wearing new foreign clothes, locals disappearing without explanation.

Meanwhile, Mr. Pete sang to himself to the tunes on his CD player, stared stiff and beady-eyed at passers-by, raked, then stopped, raked and stopped and then seemed to dissolve into the beach along with the sun late in the afternoon, only to show up in the morning at the same spot and at the same time the hotel and resort workers started their daily jobs of raking the garbage and sea grass off the beach. Mr. Pete imitates their moves: rake, take a break, rake some more, place the garbage in heavy-duty black plastic bags and pile them up ready for daily beach garbage pickup. Mr. Pete even used his own garbage bags, filled them and neatly stacked them. It all means more work for Ricky and Vernon, the village beach garbage guys, and they don't like it. *Wi gat lat to do. Nof! Wi nuh pick up afta dis tourist*, Vernon said in a gruff tone that belied a feeling he has about the place, the tourists, and a disquieting everyday life that seems all the more unmanageable as it becomes less recognizable but as more and more garbage from who knows where drifts on to the beach and accumulates like the stories about tourists drift in and pile up creating more confusion, cluttered conjecture and edginess: life feeling more and more impossible as it becomes more and more imponderable.

Mr. Pete is like one big pressure point conjuring a bundle of ambiguous images and stories, a strange and crazy tourist creating more work projects for locals (not a happy predicament), but doing a pretty good job, in the opinions of others, of keeping the place clean and setting a nice example, while also putting some locals who aren't as conscientious about their own garbage details to shame and so making them furious and embarrassed, the subjects of light local ridicule. Vernon wants to give him a beating and chase him off. Mr. Pete made Vernon and others edgy and aggressive, and that jumpiness added to the twitchy intensity of the jumpiness locals now feel around the place anyway. And that's just one line of flight, one potentializing force in the act of unfolding, territorializing onto some plateau of uneasy nervous public feelings that rub up harshly against tourist-induced images of the place as a pleasure world.

As such, Mr. Pete is a mixed force of affective excess, mysterious as he is nonsensical. As such he is an example, ". . . a singularity, a disjunctive self-inclusion, a belonging to itself that is simultaneously an extendibility to everything else with which [he] might be connected" (Massumi 2002a: 17–18). This assemblage of mystery and nonsense, dread and giddiness fashioned out of an ambiguous and odd assortment of practices, tools, and feelings, shocks many locals into nervously taking positions on tourists and tourism in general, rethinking priorities, and making decisions about tourists and other things for which there is no motivation or reason to do so more than an urgent, tetchy network of demanding and troubling feelings that are making life impossible.

Locals like Miss Gloria and Mr. Richie and their neighbors in Wallaceville made sense of Mr. Pete's reappearance through their stories, not by constructing an explanation for his appearance and disappearance or his odd behavior, but by offering accounts of his mysterious traces and effects and the nervous impacts conjured out of contact with him. And the stories piled up like shipwrecks on the reef, rocked by swirling waves of telling and retelling as the talk formed a tidal rush of dramatic and excessive images and forces that overwhelmed the merely referential and meaningful. Mr. Pete as a body without an image flashed up uncontained by meaning onto the Wallaceville shoreline. Story subjects, objects, and events became performers in a spectacle that exceeded linear reason and the discipline of cause and effect, truth and lie. As such Mr. Pete is yet another "bad example," a material singularity, an affective intensity, a force of discordant tendencies, unfinished but becoming *somet'ing*.

The power of the storytelling that focused on a strange man on the beach was that it drew listeners and watchers into a space of tense and lingering forces, some seen, others unseen, an affecting presence, a cultural poetics in the act of making something of itself. Mr. Pete became an act of creative contagion (Massumi 2002a: 19), a troubling state of suspense and suspension that haunted the place and its people, pulling them up short, and, that lingered as a troubled impulse struggling to "make sense" and make something of things more generally.

It was the impulse to make something sensible of circumstances and events by fashioning stories about Mr. Pete that turned the sight of him and his sudden disappearance one day into a tactile force. Wallaceville is a place of impacts, rapidly transforming into a spectacle of some homogenized, global, dreamworld, adventure pirate Paradise, as if by some horrid and seductive mimetic Disney magic, before everyone's very eyes, connecting locals with the global currents and flows of capital, information, people, and culture. The arresting presence of Mr. Pete entered local senses, lodging there, growing as an intensity, forming into a state of nervous suspense filled with resonance. His presence and disappearance figures the forces or intensities of what Gilles Deleuze calls affect, the double-sidedness of things where, as Brian Massumi (2002a) explains, the virtual meets the actual, and where what "matters," as a materialization of local life in the making, is the permeable edge of potentiality itself.

Looking for something positive, some sign or something meaningful about everyday life in Wallaceville with Mr. Pete in it, locals started to realize that the only thing that might be possible is some lingering, awful disclosure that acts as some anxious, unspeakable incomposability, an undecidable, crazy force of becoming, an emergent vitality that quite literally charges up the place by the sheer presence of Mr. Pete, making life impossible. Far from anything forming as named "feelings" or "emotions" fashioned out of some representational discourse or some known subject position, such emergent vitalities take shape in the surge of intensity itself as an emotion before it is actually named as such and thus placed as part of an established discourse, a moral narrative, or an ideology. Vitalities intensify at random in fleeting gestures of affect, if only for a moment, before they become folded into a normative system. The point is to evoke the vitality of things in their movement, at the moment of their becoming. More compelling than a linear narrative and more restive, multiple, edgy and unpredictable than a representation, these fractious vitalities are constitutive events or acts that animate and literally compose materially as forces of emergence.

Rapture and the Raelians

And if Mr. Pete were not enough to conjure a state of nervous disjunction, an unregistered difference in the act of its emergence, a charged up flow that makes Wallaceville feel jumpy, aggressive, and unsteady, as if life were becoming impossible, there were other tourist impacts that equally took the place by surprise, caught it short, stressed it out, and made it crazy. With Mr. Pete, public culture, and the incipient structure of feelings out of which it was being fashioned in Wallaceville at the time, ricocheted from one crazy moment of impact to another as if time were a network of punctuated moments of rupture that held a charge in the act of articulating something. But what?

Take Bob, a retired evangelical minister from a small coal camp in West Virginia, who, along with his wife Diane, owned and operated The Beachfront Inn and Chapel until it went broke early in 2008 and they sold out to "some big developer." Just who, he wouldn't say. Bob and Diane are gone now, while their resort sits rotting. It's a haunting ruin that most people stay clear of. *It gives everyone the creeps*, Richie says. He heard that some tourist was murdered there a few years ago, while squatting on the land. But while others have heard the same story, no one is really sure and there is no public record of a body or official news of the event. Such an obvious absence conjures another anxious presence that grows alarming as it touches onto Bob's intimate history with the place. Mr. Richie wouldn't have even thought about any of this if it had not been for Mr. Pete's active presence. But he insisted that I meet Bob and Diane.

Bob and Diane moved in to a little resort just north of Wallaceville in 1998. Bob felt commanded by powerful spiritual signs that came to him in the form of the Lord's voice after a series of intense "prayer appeals." But there were other signs, too, like the time he saw the face of Jesus in the clouds gathering above the bit of beach just in front of his place while swimming there during a holiday. That was his sign to buy. That's when he knew he was *taking the Lord's path*.

Bob and Diane bought the resort and transformed it into a beautiful little place. He successfully sold "Rapture tourism" for several years and made a *small fortune*, enough to think about expanding. But while he and Diane weathered the Flood that hit Wallaceville with such a devastating force in 2001 that it left nothing but heartache and rubble behind, his resort didn't, and he spent much of his nest egg on repairs. And the guests returned, *praise the Lord*, to spend a week or two at a time with Bob and Diane praying and scanning for signs of the rapture.

Bob's guests spent their time watching for special wave patterns, scanning Fox News 24/7, and praying. That was their shared evangelical mission. There is an art to all of this that Bob feels he must share with his guests. He got the idea about wave patterns while reading from the Book of Revelations. Bob said Revelations is really the blueprint, the book of apocalyptic signs that in 1998 he could no longer afford to ignore. Bob figures that the signal for the Rapture will be found in strange weather that will be preceded by unique wave patterns, as the sea prepares itself. The Rapture, he calculates, is very close at hand, so it is important to be vigilant. It's a matter of logic. The Flood was only the precursor to the big event. Fox News stories, he figured out on his own with no help from the Bible, would cluster into a pattern of messages, too. News and the waves were parallel signal systems that he and his guests turned into a "detection machinery."

These tourists were not distracted by their beautiful surroundings. They came to scan and pray. The locals and the nay-saying "hard drinking" expat "philistines" that lived around them started to compare Bob and his visitors to another experiment in tropical evangelical work and worship in Jonestown, Guyana. The Jimmy Jones cult effect, the commanding voice of the Lord, the dreams of a final Paradise, a return to Eden, martyrdom, Koolaid and the glorious ending of the world in the Rapture: all a Belizean tourist-made line of flight pushing up against a growing yet diffused structure of village feelings (public sensibilities in chaos, a place and its people and guests growing oddly amused, aggressive, unstable, and strange) a tension so palpable that it also served to make life increasingly impossible. That's when Richie asked one day: *Why do we get all the crazies?*

And before anyone could give Richie a good answer, an answer of sorts presented itself.

In 2009 the Raelians were serious about establishing an embassy in Belize, a welcome center for beings from another planet, the uber-tourists, the ones who created life on Earth in the first place, and who now wish to return to visit and take a tour of their experiment in cloning. What better place than

Figure 2.1 *Amandala* newspaper article announcing the Raelians. (Adele Ramos, "Raelians Want an Embassy in Belize," *Amandala*, 27 March 2009.)

Belize? "It now can live up to its name. Belize is an ancient and sacred place." Bernard Lamarche explains on Channel 7 TV news, on April Fool's Day 2009 no less, while standing next to nervous looking government officials. The Raelians want to give Belize tourism a whole new look and purpose, and spin it into another dimension. *To boldly go where no one has gone before?*

Richie asks. And the suggestion stops him short, and sadly he begins to wonder what's happening to the little nation he loves so much. The Raelian offer morphs the state of imperial collapse (the state under chaotic contemporary conditions as international capital mysteriously disappears and reappears from the center of power, partially through the efforts of schemers and dreamers whose ruthless cunning has shaken the Belize economy) into panic mode so that a solution like the Raelian offer begins to sound like its own salvation. It's how a mysterious, cloning culture of Raelian enterprise suddenly appears on the margins of empire and of the possible, to save the day in Belize. It's a new take and tale on the imperial magic trick that conjures as much giddiness as it does uneasy, impossible expectation.

They say that the money to be generated from the multi-million-dollar Raelian embassy/tourist center would be in the hundreds of millions of dollars and attract more than fifty thousand tourists a year. "That's better than the cruise ships to which Belize has attached itself ball and chain," the Minister of Tourism, Civil Aviation, and Culture said on the Love FM *Morning Show*, his voice a strange mixture of authority, confusion, excitement, and hopelessness. With that kind of foreign currency in the assets column of its national development plan, Belize may stand to win big time. But Belizeans have heard it all before: the impossible promises, the unfulfilled expectations, the failed projects. *Who knows, a welcome center for aliens might catch on. It can't hurt. We'll have spacemen for you to study Ken*, Richie quips. *We'll call it super-natural Belize*, he chuckles sadly.

Empire eclipsed by the outer space parent aliens, tourism in the grandest of styles conjuring another odd and nervous tourist moment in an impossible Paradise. Here the productive forces of global capital are at work on a new presence, a new now, a future history of tourism construction sites for outer space guests along with whole new tourism infrastructures of resorts, casinos, spas, golf courses, and ecotourism technologies, all sovereign enterprises that rub up against the ruins of the rapture, a new after-history of future decay and world destruction. And in the gap, in the present, in the now, the impossible job of keeping fore-history and after-history from collapsing in on each other.

Richie's New Tree Project

Over the last few years, before his untimely death, Richie had cultivated another relationship with trees. He named a big flamboyant tree in his yard the "Tree of Truth" and put a sign on it. Everyone thinks it was the tree he used to hide the money he found. But if it is, Richie isn't saying. That means everyone has a theory and it keeps local interests up and everyone in the loop. Richie does say that he uses the tree for "teaching purposes." That means he sits under the Tree of Truth with unsuspecting tourists who are seduced by his antics and wicked sense of humor. They buy the beer while Richie tells the stories. They end up very drunk and short on cash. But that's how tourists find local wisdom, Richie says. He says that the tourists wake up feeling "enlightened" if not a lot worse for wear, wondering what happened, feeling a sense of anxiety mixed with relief that they are still in one piece but happy that they got to share an intimate moment with the locals. Shared wisdom all around, that's the way Richie likes it. It's a world of shared banalities masking as local flavor that can be a basis of sociality or an exhausting enervation, or simply just something else to do. Little moments of contact are felt as pleasures and warning signs, as exotic intoxications and repetitious daily routine. It's an odd ordinary that composes itself out of seductions, intoxicating encounters, local need for distractions, shapeshifting solutions of a sort that help shake the drudgery of the everyday with something, anything, else to do, in a place that is transforming into a monstrous and banal tourist pleasure space unlike anything any local could have ever imagined. It's a Wallaceville where disparate events and sensations come together to form an odd ordinary, the repetition of which leaves a residue like a habit—a living cliché and another moment in which life becomes impossible.

Apocalyptic Dreamworlds

Here we have a set of intersecting and entangled moments that collectively fold and unfold into a sense of strange goings on in Wallaceville and that mark life there as impossible. The uncanny sensation of half-understood invisible apocalyptic or unexpected and contingent forces, specters and spectacles, powerfully populate the place and possess it with strange new spirits. Private lives and public worlds getting their wires crossed and snagged up on each other. Dreamworld and catastrophe, success and failure, prosperity and collapse, the Universe and the local, exist alongside each other, inextricably tied as immanent to each other's details and making (Buck-Morss 2000). Here, an incipient structure of public feelings begins to form along the lines of apocalyptic discourses that take the shape of the Rapture and the Raelians or in the form of mysterious flows of money or sheer boredom; it throws itself together through affective forces, energetic incitements as much as material signifiers. This set of forces of public feelings is life becoming impossible. It is an affective becoming that resonates as an included disjunction, vibrating with tension that lodges in the body, histories executed through the body in lines of tension and relaxation as happenings, as things happening. Whether these affects are feared, seductive, romanticized, subdued or unleashed, they always point to a generative immanence lodged in things as they take shape in the surge of intensifications as moments of vital impact.

The Rub

There is not a day that goes by that someone in Wallaceville does not say something about how life is becoming impossible. Impossibly smooth and beautiful, impossibly cruel and corrupt, impossibly laidback, strange, and seductive, impossibly transformed socially and ecologically, impossibly out of control and violent, impossible to live impossible not to live. Through this set of stories I track the impacts and intensities of life becoming impossible and the formations of bodily sensations that open onto the affective intensities of life becoming impossible, as a scene of immanent forces folding into an assemblage of public feelings, public culture in the act of its becoming. I am trying to track the troubling state of suspense and suspension that haunts the place and its people and that lingers like a jumpy, chaotic, and creepy impulse trying to "make sense" of things that come into view as habit, shock, intensity, resonance, or resistance (after Stewart 2007: 1–7). Lives in the gaps, or the interstices of this contact zone of encounters, the middle of things, throw themselves together as event (movement) and as sensation (affect), something becoming, some incorporeal materialism, a disjunctive encompassment, dreamily inhabitable but exhausting, a tropical dreamworld and an uncanny ordinary. To reiterate: Examining moments of encounter and lingering in the impacts of life becoming impossible means tracking free-floating affective agitations and sites of collective feelings as the movements of emergent and potential emotional forces coming into play in this new state of emergency taking shape as neoliberal exception, on the edge of global empire, in Belize.

But it is here on this uneven terrain between what can be imagined as possible and what may be beyond the scope of the possible altogether, the impossible, that forces of sensation begin to instantiate. Impossibility, or the potential of rupture between the chaos of a world of the possible and what acts beyond it, a moment without end, an unspeakable, unrealizable, that which escapes the grid of intelligibility, has no horizon, a passage not a presence, a becoming, a becoming impossible, that which I cannot conceive yet nevertheless reach toward (Manning 2007), an expressive fragility in the making of some condition, good or bad on a beach, becoming otherwise, in Belize.

Chapter 3
Richie's Tourists

Happy Hour Hank

Some people are wise, Richie says impishly, *and some people are otherwise. That's what my daddy told me.* It's a charming bit of local wisdom that Richie loves to share. In this case his words were directed at the "otherwise" "Happy Hour Hank" who suddenly appears at Richie's table at The Sundowner during happy hour and tries to take a seat. Hank is "Happy Hour Hank" because most villagers know that his happy hour starts around lunch each day. It's just another version of an ordinary expat daily routine that isn't his alone, a rhythm of movement that leads from the safely secured compound he now calls home, protected by two vicious dogs that the neighborhood detests and complains about all the time, and into village life for shopping, visiting other expats, for community news gathering and the usual round of gossip about the locals, and about who, or whose place, just got robbed, who is fighting with whom, threats, special events, and why you can't call the police about crime because they are too lazy or corrupt to do anything about it. The thief could be almost any drug-addled local from up the main road, take your pick which one it is, they say. This all leads to another round of drinks for the thirsty group, struggling with the glaring intensity of the midday sun, and then another round, and another, and then on to the next bar. In Hank's case the rhythm of his daily routine is his way to keep on course, busy and informed: precision home maintenance, visiting, talking up tourists and other expats like him, talking to the locals that will still talk to him, watching tourist boats coming and going from the Wallaceville pier or that anchor off the caye, getting the "daily buzz on."

But Hank is not alone as an expat who has learned the magic of extending happiness to fend off the extreme and impossible boredom mixed with the dread and fear that Paradise has generated over the years of trying to live and live up to a tropical dream. Once the happy, tropical dreamworld seduction of the place (the place that was Hank's initial stimulus for retirement, with the help of a reasonable pension from his old, boarded-up, central Kentucky industrial job) morphs into nightmare disappointment and drudgery, suspicion, fear and depression, where do you find getaway happiness? Once you have fixed up the house and yard with the latest security systems, air-conditioning, and the new garden and a new dock, after you have tried all the local therapeutic systems and ducked another skin cancer scare, after you have tried to help out in the community, only to be rebuffed or ignored by the locals who steal your boat and motor and laugh at you behind your back for all your social earnestness and control issues, after you have started a boat and motor repair business that doesn't get off the ground because locals figure you are "colonizing" possible village initiatives and nixing economic futures, then what? Hank tried everything to keep busy and "on course" for the first couple of years of the seven he has lived in Wallaceville, only to have his seductions rub him to the bone, making his life in Paradise feel next to impossible.

But Hank figured it out. "Be like the natives, more or less," he says. Now he lays low. He found a local Mestizo woman, Rita, with whom he now lives off and on, but she tells me in confidence that no one in the village likes Hank and that she has so many doubts about living with him. Even so, she stays on, but she is not entirely sure why when he can be such trouble and an embarrassment. Rita thinks Hank is an alcoholic. *But then again who isn't,* she complains. Hank doesn't see how angry he is, of course. He says he's just standing up for himself after being seriously ripped off by no-account locals one too many times. He won't take any more *rass* about his drinking either. He knows a lot of stuff about locals, enough, he often threatens, to get them into big trouble and that aggravates everyone and makes him a satisfying target of derision and steady low-intensity contempt. So Hank has shifted focus, adopting a daily routine of an afternoon tour down the beach to the lagoon and to various bars for drinks, tokes, talk, and amusement. Anything to relieve the relentless boredom that builds up and thickens into a frustrating undercurrent of loneliness and a distrustfulness that, more often than not, dangerously drifts into self-abuse and despair, and the constant feeling of being trapped in some horrid, perverse pleasure world with no exits, no escape, other than a return to Kentucky, something Happy Hour Hank says he can't manage to get his head around right now.

Hank reaches Richie's table at The Sundowner about the same time the rest of us do, in time for our daily, late afternoon Paradise drinks and tokes. But he's way ahead of us, positively lit up, smiling but edgy, and on the verge of who knows what. Hank seems both crazy-alert and loud, obnoxious and funny, angry and sad, sappy stern and maudlin all at the same time. He instantly begins to heat things up just as the sun begins to set gently over the back of the lagoon to cool the rest of us down. Hank tries to tell us a fishing story, but he's nearly incoherent and the story sounds like a jumble of themes and dissonant sounds that do nothing more than pump up the nervous pressure of the situation. And that's when Richie steps in with his own *rass*, encouraging Hank's dissembling, and this only adds another nervous edge to another early evening tourist bar scene, generating a shiver of dread and panic that takes on the strange texture of an unpredictable, unfolding moment that looks like it is sure to head off the rails. Everyone at the table can feel it.

What is emergent is a feeling, something like the feeling you get when a super storm is imminent, your skin tingles with the sudden change in atmospheric pressure, there emerges a darkening ominous presence, and you can feel it coming, smell it coming. *Hank, why do you carry so much baggage? Get smart. Pack up. Go home*, Richie spits, all the while smiling and looking about the table for support. From the mixed expressions on Hank's red face it's clear that he can't figure out if Richie's serious or not. Hank just stands there swaying: blockage, bad energy buildup. He looks as though he may feel an angry reply of sorts grow in him, but it's five in the afternoon on a very hot day in a busy bar full of sailors, fishermen, and tourists and it's obvious that he's in no condition to put his feelings into words. Instead, his feelings (some crazy assemblage of bewilderment, anger, embarrassment, exhaustion, dissipation, a history of disappointments) are yet unnamed. They are emergent sensations that begin to throw themselves together as the scene composes, as an atmospheric pressure builds to become an attunement, an alerted sense that something is happening, that something matters (Stewart 2010). That's when Hank suddenly snaps.

Seven years and many hours of trying to map out his routes and routines of happiness in Wallaceville and the cayes, almost all grumbling disappointments, have generated a sort of sad sixth sense for Hank as he still tries to settle into the village. Right now it's a confused awareness of deep anger mixed with frustration, sadness, incredulity, humiliation, and awkwardness, and it all begins to mix and reach out from a drunken dimness and move around Richie's table as Richie's *rass* shakes Hank into some confused state of alertness. It's too much. He tries to push the table into Richie's face, but he struggles unsuccessfully to lift it. He looks at me as if I should understand and help. I'm no help. Hank calls me a traitor. He calls me a spy and tells everyone that I can't be trusted, that I'm working for the CIA and the FBI, maybe Homeland Security. Then he falls over, gets up, and falls over again. Some of us try to help Hank up while pleading with him to be reasonable, not to take things so seriously, *Richie's just bullshitting. It's rass man, you know how it is. It's shit. Take it easy. Have another beer. Here, have a toke*, only to be roughly pushed aside. Hank now looks ferocious and turns on us, to leave maybe? To fight? No one is sure, but everyone is now intensely edgy, and a hot sudden stillness reaches across the place. Instead, Hank begins a lonely stumble out of The Sundowner, and in the dimming light, trips onto the sand path that may or may not lead him home. It is impossible to be sure that he can take another step, yet slowly he does, but barely.

The tourists who watch Hank struggling consider the scene disturbing: Paradise to excess, a man who doesn't know his limits or when to say no, or how to behave. But they don't know the half of it. Those who think that way don't understand what it is like to live in a place like this. It is Hank today but just as easily it could be you next and that means that expats do not spend more time than they need to being critical. Hank's lesson is a local and serious one about what happens when you don't stick to your routine or you stick to it to excess and so you don't keep in reasonable touch with reality while the seductions and frustrations of Paradise constantly work on you. So Rita and other locals take it all in stride, sympathetically, even though many of them don't like Hank. Again, lots of locals have been right there, where Hank is right now. Most who live in the village have had their embarrassing moments more than once or twice. It's not the first time this has happened, and it won't be the last, for any of them. So it's not so much a local criticism that spreads through The Sundowner as it is a lively lesson in local living. But still, everyone laughs, Richie's laugh maybe the sharpest.

Stripped of all but the urgent need to keep moving, pretty much on "autopilot," Hank feels his way through the growing darkness, each unsure step adds intensity to things even if things never seem to help him advance his way home or move his life forward in any good way. From their window slats, some of Rita's family watch Hank's near-collapse and phone her to come and get him. Rita finds him on the side of the road, in the dark, numbed by the day and its events. It's that time of the day, the moment of its complete dissolve for Hank, now barely a pulse, bereft of passion, an irrational expenditure that is an emergent threshold of chaos. This is all about how things come to matter— and through what qualities, rhythms, forces, sensations, relations and movements—to help world a jumpy, nervous, impossible Paradise. Rita patiently walks Hank home.

In the meantime none of us is very happy with Richie for having initiated what was Happy Hour Hank's inevitable collapse, and this is not the first time he's done this. He gleefully picks on Hank and has for years. In fact, Hank refuses to talk to Richie any longer or frequent The Sundowner. A few of us are concerned that if Richie keeps it up Hank will turn into vicious trouble in a click. As it is, he's a time bomb of potential danger with a trigger-personality, that makes us all anxious whenever we meet up with Hank, even when he seems happy and sober. No one's very happy that Richie took advantage of Hank's condition and broke up an otherwise pleasant vibe. Nevertheless, that Richie refuses to take Hank's *rass* demonstrates just how brave he is. For Richie such control is a demonstration of local "smarts" and power and that's why everyone calls Richie the local psychiatrist. Cedar, a steady bar tender at The Sundowner, calls him a "witch doctor," others call him the "pirate-psychologist." After all, who is brave, smart, or stupid enough to provoke disturbed expats, drunken sailors, local crooks, national land and drug dealers alike and in public, into spilling their secrets. For some, contact with Richie is a local but powerful enchantment of sorts that more than matches the discharging power of the Holy Ghost and the creeping growth of the evangelical preachers in Belize, or the influential persuasion of the big politicians, or the "good life" seductions of the tourist resort entrepreneurs, or the alarmist discourse of the burgeoning environmental NGO sector, all of whom act as credible substitutes for the "old days" and who are responsible for hollowing out the Belize state and civil society and reassembling it as a state of exception (see Piot 2010).

But by now Richie is feeling the effects of his happy hour imbibing too. He's not paying any attention to us and he quickly moves on to a table of tourists who recognize him and offer to buy him drinks. *We Creoles know what to say*, Richie shrugs. *We just don't know when to say it.* The statement is directed at me and he gives me a "sorry" look, but the tourists pick up on what they consider to be another example of Richie's locally inflected wit and they laugh. They want Richie to blow up the Paradise balloon again with some friendly "native" contact. Sitting with Richie, the tourists are secure in the sense that things are just how they should be. That's when Richie meets tourist visitors and that's when he invites them back to his place to sit under his beachside "Tree of Truth."

The Richie Touch

Mr. Richie is seductive like a sudden and surprising touch that jolts straight through the small of your back and into your thoughts, a feeling without a name. The impact of Richie's seduction stays with you in the troubling way a gossipy conversation does, when you have gnawed at someone else's negative qualities or secrets but end up revealing your own pettiness and insecurities. Richie calls that "baggage." Or like the crazy way a lingering scent remains in the air, unidentified, thick and heavy as it rides the waves of heavy June humidity on the coastal cayes in Belize, disturbingly seductive for the immediate effect it has on you and what gets conjured by scent sensations. You draw the smell in and an impulse grows intense, only to be transfixed by the impact of the sensations activated by the flash of images and the tug of feelings triggered with each breath you take: an unattached, floating sensation that has an impact, an active generative force of something in a state of emergence. But that's how a seduction works, as a moment of generative emergence, a sensation unfolding, something still not assembled as a concept; things, instead, in a shifting state, renegade, opportunistic, haphazard, unsteady, and unfinished things in the eventfulness of some surprising realization (Stewart 2007: 2).

It is contact with Richie through his funny stories, verbal antics, and perplexing performances that index some seductive unfolding of an emergent generative force of the "local," which over the past fifteen years has quietly moved Wallaceville into the Caribbean tourist scene as a "go to" destination. And there is not a day that goes by that some tourist somewhere in the village doesn't say something about contact with Richie. A middle-aged English sailboat tourist told me that "everyone's just wild about Richie." She and her husband find him "engaging," "knowledgeable," and "amusing, rotten teeth, scraggly beard and all. He's like a cute pirate," she bubbles. She knows because she and her partner sail the Caribbean around Honduras, Guatemala, and Belize as part of their retirement project and have developed quite "an eye for tropical culture and adventure." Richie has lived most of his life in Wallaceville, and over the years he has watched his "sweet little beachfront village" go "crazy" for, and get "roughed up" by, international tourism. A fisherman as a younger man, then one of the very first village tour guides and operators, now, in 2014, at sixty, he has become a popular "man about town," a "Creole Cassanova," (a tourist's description), and a modest tourist entrepreneur who rents out beachfront cabanas while dispensing "native" insights about "Creole life and experience" to short- and long-term visitors alike. Often such dispensing occurs around Richie's table at The Sundowner or under Richie's tree grove. *It's a living*, he laughs, commenting on his role as a "performing" local.

Richie often invites visitors to sit with him and that's when he beguiles and ensnares them with his antics, ribald jokes, mostly good-natured abuse, and his trademark laidback manner of engagement that acts like a local and seductive "ready-made" for tourists. Richie's smooth and relaxed "don't worry, be happy" performance activates tactile images of a generic carefree "native" life in Paradise. It is seductive, pretty much because Richie is seductive, and that's just the way he likes it. His image, his stories, and his personality combine to become a force of copy/contact, a force of mimetic excess that tourists enthusiastically go for, or not (Taussig 1993).

Richie's Expressivity as Bad Example

It's not a matter of whether Richie believes his own stories or performances, nor is it a matter of dissecting the representational politics of his routines, as much as it is a matter of attending to the charged up tendencies of abjection or seduction that are fashioned out of moments of "Richie contact" and that together compose themselves as a "consistency," a dense entanglement of affect, attention, and matter (Grosz 2008), "slowing the indeterminate chaos of sensations enough to extract something from them that is not so much useful or meaningful as intensifying," a potentiality and an event, an expression; "affective dynamized forces" rather than a system of signifying images with a context: in short, a force field of contact (Grosz 2008: 3). I say a force field of contact rather than use the more recognized concept of the "contact zone" (see Pratt 1992) in order to distinguish the way the zone is often used as a social description of liminal creativity, as a space of identifiable affects, a sociocultural zone of mobilizations that engage given political and economic infrastructures not from one side or the other (tourist or local) but from their merging. Force fields of contact may include all of this in that they are interstitial and engaged, but they are more than that too; they are more like different edges of an interface. As both actual and imaginative fields they are a milieu and a trajectory that enact new infrastructures of difference, new modes of relatedness that affirm, augment, and keep open the spaces of dissolution and irruption, the charged up potentialities that index contact and further facilitates the multiple occasions of their collective presencing (see Haraway 2008). "The trajectory," Deleuze says (1997: 61), "merges not only the subjectivity of those who travel through a milieu, but also with the subjectivity of the milieu itself, insofar as it is reflected in those who travel through it."

Fields of contact are autonomous zones of virtual-actual circuits (see Massumi 2002b: 1–45), at once social and material, affectively infused intensities and trajectories, by which I mean co-constituting movements through milieus as enactments of desire, need, curiosity, or simply attempts to find room to maneuver and breathe in the entrenched political and economic system of a tourist state in Belize. So too, tracking Richie through fields of contact is more than simply treating him as a constituted subject within a social formation revealed through his experience. Not only do the fields of contact in Wallaceville and in Belize "leak," not only are they incomplete and so unable to constitute "subjects" like Richie in any determinant sense. When Richie speaks and acts in concert with others, he does more than reveal an experience as it has been constituted for him. He also performs a desire toward becoming that is much more than an experience can govern and control. Richie enacts a worlding as an assemblage of sand, water, atmospheric pressures of beachside temperaments and temperature, tourists and local bodies, palm trees, color, and climate as they touch upon and become vibrant in contact.

In the process of being lured by a force field of Richie-contact, as a milieu of transiting social, material, and psychic strata, a transitional immediacy of real relations of desire, opportunities and challenges present themselves not only to track new forms of sociality but to enact a sociality that is an infinitely open process of collective self-elaboration. Such a sociality composes and consists, assembles and reassembles, in ways that are incommensurate with the telos and demands of the imperial prerogatives of tourist capitalism as they are enacted on the frayed margins of empire or to any extant political-economic formation of the good-life plans of Belize tourism (Massumi 2002b; Manning 2007; c.f. Povinelli 2006 and Dave 2011). For Richie it's a matter of attending to what's happening, sensing out, attuning, accreting attachments and detachments, differences and indifferences, wins and losses and proliferating possibilities in a conflicted tourist dreamworld. For me it is a matter of attending to the attunements to such a milieu of contact sensations. It's a matter of taking "a step sideways into what normally gets stepped over" (Stewart 2008: 4) in the social analysis of tourism in general and in Belize tourism especially. It's a matter of being right where you are only more intensely, in the eventfulness of Richie's performative expression, in Wallaceville, in the heat of the Caribbean, in a bustling tourist destination dappled in that official all-too-blue that credentializes any tropical beach scene on the margins of empire, and to pause and wonder what might matter in singular moments of contact with him, attuning to the alerted sense that something is happening, and sensing out whatever it is: seductive contact as a generative, compositional worlding.

Tracking such acts of composing means dropping an insistence on analytical binaries like the material and the representational. It means not holding concepts up for evaluation as good or bad, or as subject or object relations. It is not trying to present a critical analysis of touristic contacts with Richie as an instance of Belize social change under the several characteristics of global empire that by now are habitually employed to define such processes. Rather, it is sensing and evoking the capacity of a field of incommensurate forces that somehow hang together as a consistency, a complicated intimacy of things that matter because they are thrown together into things happening—little moments, big scenes, nervous encounters, uncertain intensities—as they pick up texture and move through bodies as expressions. Mine is a paratactical writing that lingers in the milieus of edgy contact where bodies are fashioned by repetitions of forceful sensations as a sensual affirmation, for better or for worse. So I take up the precariousness of writing and the potentials of what Deleuze (1997: 104–5) calls a "minor literature." This requires more of the subversive's creativity than the social scientist's methods of description, born in part in a willingness to cross planes of reference without regard to received expectations that translation and order are supposed to control difference. This is what I am trying to create with my writing: paratactical expressions of Richie on a plane of becoming, or better, Richie worldings, a milieu of uncontainable movements that form as a contact situation, some unstable version of events becoming one more instantiation that stands along with so many others in a nonhierarchical extension of forces as a field of composition, a consistency (Deleuze 2001: 25–31).

Thus it is to a consistency and the various forces of its expression that are in and of the body, both sensing and sensual that I turn my attention. Massumi (2002a: xvii) argues that the "force of expression . . . strikes the body first, directly and unmediatedly. It passes transformatively through the flesh before being instantiated in subject-positions subsumed by a system of power." That means that its effect is one of ". . . differing. The body, fresh in the throes of expression, incarnates not an already-formed system but a modification—a change. Expression is an event." It is to world-making entanglements that contact with Richie, and those with whom he makes contact, provides the opening for an event, for eventfulness. Expressions, then, are not representations or signifying illustrations, instead they are charged up, co-constituting, material-semiotic entanglements of bodies and meanings in interaction, in encounter, sometimes to a body's advantage, sometimes not (see Haraway 2008). So bodies—Richie's, the tourists, and mine—are vectors of emergence that generate virtual embodiments, co-constituting force fields of contact we can only reach toward: an expressivity as co-constituting entanglement.

Again, the events of Richie contact are actively generative, a state of emergence that I track by side stepping or drifting along with things, attending to the crazy vitalities of free-floating allure, bodies (agitated, delighted, or enduring), and to contingent moments of collective animation and exhilaration that compose the atmospheres of life in a Belize tourist village (see Stewart 2007). It is about how expressions of life in Wallaceville lure, sustain, build, or suddenly splinter, collapse, or explode. It is to how such things compose a present that can seduce and coerce fugitive or shifting or funny or pushy or hostile forces into forms that actualize or qualify a texture and density that can be felt, used, imagined, or not. Attention here is to these states of emergence and their moments of flow, impasse, retreat, incongruity, confusion, and exhaustion in the face of a tropicalized Paradise. The stories I create about Richie track lines of intensity as they emerge in ambivalent moments of delirium, or become consequential in local everyday feelings about things, or in common sensibilities and uncommon moments of shock or absurdity. How do locals, expats, and tourists become actively charged up in and by their expressiveness, by the rush and pitch of things in the making, wild and twisting assemblages of impacts as an emergent consistency?

Tourism, the latest instrument of Belize economic development, has transformed Wallaceville, and caye life more generally, into a seductive "Caribbean pleasure-pirate hideaway destination." Everyone in the village knows things are "happening," that they are now oddly, but deeply, into something.

That means everyone is in a permanently edgy state, on edge on the edges of empire. Ask anyone and they will tell you about how life is becoming impossible in Belize. There are the stories about the winners and losers in local land deals that have literally possessed some, dispossessed others, or disposed of them altogether. The stories of jealousies, marriage breakups, drug deals, the sudden appearance of new fishing equipment and tourist boats and motors, cars, clothes, houses, and friends from who knows where, and a whole new local population of expats or cruise ship tourists who will not stop arriving or pushing store prices up, or demanding new services.

Stories encourage more stories, odd or stilted conversation among old friends becomes an exercise in divining strange successes or failures, each conjuring new arrangements and assortments of secrets or divining winning lottery numbers. People all of a sudden start going crazy or dying. A rash of fires one night just about burned down that whole section of the village where the old wooden houses, hotels, and restaurants stand as withered and dry as tinder. The nervous intensity was thick enough that night, and for months later. You could smell it on the burnt remains of the places that were torched, throughout the village and the cayes. Locals always attend to what's happening and this in and of itself materializes in this shared eerie and unpredictable sense of life becoming impossible, a shared intensity caught in the nervous impulse now to keep secrets safe, guard your talk, stay on guard, look out for more gossip about eccentric strangers, strange friends. Bodies are forever on alert in Wallaceville in an awful watch for the next kid to take a baseball bat to his best friend for stealing his hat or his girlfriend, for the next big expat robbery or house invasion. But daily expat demands are insufferable many local Creoles say, and many of them deserve what they get. Or locals are on alert for the tourists, for the sex, maybe, or for a trip to Europe if they really luck out, or maybe just some extra money. So the tourist came. The expats came too. CNN came. The drug dealers came. The environmentalist NGOs came to save the reef, the jaguars, and the jungle. Money came. Local girls learned how to dress up. They joined beauty classes, watched "American Idol," and tried out for the annual bikini contest to become "Miss Belize" or a Belikin Beer calendar girl or at least a waiter at a local tourist bar. It is all about the unruly possibilities of living in a place of social confusion, despair, potential, and hope.

Local life in Richie's Wallaceville shakes violently between such phantasmagorical seductions and ever-expanding moments of condensed and alarming displacement. The only real work now is in the tourist industry and too many of those jobs are going to "outsiders," even expats, with more experience or education, or to illegals from Honduras and Guatemala who are the bare-life labor, building expat hideaway beach mansions and who fill Pastor Pat's evangelical church "Wisdom of the Ages" three times a week in order to be filled with the forces of the Holy Spirit's magical agency expressed in the language of tongues and dollars, and in the intimate details of a new personal relationship creating the possibilities for success and wealth. In short: new biopolitical life of displacement and difference folding and unfolding to the soundtrack of everyday street laughter and hustle. Meanwhile, Richie gets by comfortably on the local margins of a tourist industry growth market with the dynamism and imagination that back-talks contingency and privation with a particularly local, Creole family pride, despite the impoverished political condition of the Belize State. He has learned how to survive, he says, but the secrets of his success are his and his alone. They are the kind of secrets that I can only sense when Richie gets up to something, that's when a line of stories begins to throw itself together and in ways that urgently sound the promises, disappointments, astonishments, crazy stillness, panic, and impossibility of his everyday life. Contact with Richie, through his stories and antics as expressions, sediment his life and give it a spicy weight not unlike the steamy stillness that hangs on a languid, hot late afternoon on the coast of Belize. Contact with Richie, in other words, strangely registers the waves and pulse of public feelings in Wallaceville.

Richie is another "bad example," an expression, an emergent vitality, a singularity. So contact with him through this writing hones attention to the way that the tactility of such contact starts to take form as a composition, a consistency of materialities, sensibilities, and movements. Writing Richie contact as something emergent means slipping the system checks of ethnographic production that are the commensurate qualifications of representational thinking with a focus on self-identical objects. It means attending to the sea change of life in Wallaceville and to the intensifying atmospheres of Richie's expressivity and the fields of contact during a time of always already transitioning, a future anterior of imperial, biopolitical sovereignty (Massumi 2002a: 15; Manning 2007: 116). This is a presence that demands we not simply restate the injustices, inequalities, asymmetries of economic and political power, the discriminations, and the ameliorative economic and social development projects of the past, present, or future. Rather, we must track "presence as a transitioning" and that demands attending to both unresolved and unfinished modes of care, mutual reliances, hopes, and to the disturbing excesses, chaotic dissemblings, the unhinged pleasures and alarms, that compose a transitioning presence in Belize today.

Half-Jack

Contact with Richie, when invited to join him at his place, under his "Tree of Truth," may begin in the safe and tactful seductions of secured, generic tropicalizations, but a touch of Richie incites possible entanglements and charges up a "reaching toward" each other that exceeds the tactfulness that indexes the genial tourist and local interactions, conjuring monstrous crossovers, confusing and half-understood urges, derangements, and outrageous behaviors that are often way more than any tourist may have bargained for. Hanging out with Richie is a seductive "con-tact" moment of difference. If the "tact" of con-tact is the "delicate sense of what is fitting and proper in dealing with others, so as to avoid giving offence" (Manning 2007), and so is a body secured by a structure of vacation habits, tourist images, and local etiquette, then the "con" of con-tact, is "the reaching beyond," a contingent, potentializing force imminent in any engagement with Richie, for better or for worse. It's a "sensing beyond security," a sensing, in this case, conjuring ecstatic excess (Manning 2007: 145). Richie, the "con" man.

Half-Jack and Janet are new to the village. They are a nifty, late-fifties couple who, less than a year ago, bought an expensive condo in one of the big Wallaceville resort developments that hasn't tanked (yet) during the most recent imperial 2008 economic meltdown. Half-Jack got a "real deal," a "steal," he calls it. He did his research and had *the seller by the balls*, he barks loudly. He just sold his very successful biotech start-up, headquartered in New Jersey, that's supposed to go public. Half-Jack *made a killing*, he says. And as a venture capitalist he can now begin to work *pretty much from anywhere*. Half-Jack and Janet initially planned to hang out in Wallaceville several times a year and for several weeks at a time. But now nobody knows what their plans are.

Half-Jack is a short, loud man. You know he's in the room. He admits he's "Type A," that he's aggressive. He likes *to play the game*. Half-Jack says that he *works hard and plays hard*. Richie calls him "Half-Jack" to distinguish him from Bosun Jack and Double-Jack. Three expat Jacks whose nicknames each come with a story, or an observation. Because of Half-Jack's short, stout stature he looks to be about half the size of the very tall Double-Jack, another very recent expat bourbon-drinking homeowner, hence the names. Bosun Jack is another story altogether. The nicknames keep things clear about which Jack locals may be referring to in a conversation, but there is more to them than that. Calling Jack "Half-Jack" makes him angry. His aggressive no-nonsense business acumen and his very loud personality prove he never does anything halfway. But that's just the point of the local *rass*. Nicknames are also forces of copy/contact. They are locally invented names that form as intimate comical shorthand, indexing a field of impressions, behavior, or personal characteristics that a subject can grow into and even take advantage of in a place that values an "easy" sense of humor. Otherwise it's sometimes hard to distinguish one new white man from another.

We are sitting beachside on Richie's piece of beachfront under his grove of trees now giddy with the effects of our happy hour buzz, listening to him explain the unique characteristics of the flamboyant tree to Half-Jack and Janet. Most of us have heard the story before and stop listening to the tale about the special, even magical, characteristics of the flamboyant, about its affective presence in the village, about how it "feels happy" in the presence of community, how it needs people around for it to grow and blossom or it will wither, about how local Creole social life in Wallaceville began under a particular flamboyant, now a well-known landmark in the village. It is located on the main road of the village and has a bench built just in front of it. It doubles as the local gathering and gossip spot for those who live in the area. The flamboyant loves a good party, Richie insists, and given how Richie's party is going, his tree has to be very happy too. It is a local Creole tradition to entertain under the flamboyant, Richie says half seriously. Half-Jack and Janet are entranced by the tree decorations (small talismans, fishing equipment, a Barbie doll hung by the neck, old fishnet and floats, various sorts of boat equipment). Richie takes the opportunity to tell elaborate stories about this stuff and about local life, tailor made for the new visitors. They are eager to hear more and it's clear that they want to play a bigger part in this local scene. It's an exciting moment of contact and a legitimizing one. They know now why they made the big move to this tropical Paradise, it's for the magical, seductive buzz of moments just like this.

Half-Jack takes a wild swig from a big bottle of local One Barrel Rum then passes the bottle around, while Richie tells more stories about how tourists find "local wisdom" under the flamboyant. Half-Jack is intoxicated but Janet has passed out in a hammock leaving "party hardy" Half-Jack, the guy with the reputation for never doing anything halfway, to finally lose complete control of his senses in a moment of mattering that morphed his body into corporeal bulk that caught everyone by surprise.

With every tourist holiday there are moments of emergency, when unplanned things happen, accidental things and things not so accidental after all, like right now. At this place and moment Half-Jack spontaneously shits himself, a private moment of complete and hideous abandon made starkly public. It is the sudden then lingering stench that catches us up first. Shit is in the air. Most of us find the source. Half-Jack is sitting in his mess, on a piece of driftwood, smiling wildly, too drunk and stoned to know or understand what he has just done: defilement as a complicating enactment, an emergence of new potentials, a grotesque and shocking situation that exceeds its actuality, an uncontained materialization, a transitioning moment on the way to becoming a "situation." Half seriously, Richie asks Half-Jack if he has a medical condition. Half-Jack's smell is central to the situation and Richie's question interrupts Half-Jack in the middle of some groaning reverie, some blurred, pungent present, and he looks at Richie with a puzzled face. Richie points out the trouble. Half-Jack finally gets it and laughs hysterically. "I shit myself!" he yells too loudly. And in the full flush of a stinking moment, he gets up to do a little song and dance. But there is ambivalence. Half-Jack's dance is oddly a grotesque pleasure like some bodily protest, yet it is mixed with some dull acknowledgement of embarrassment, hilarity, shame, cheer, dread, mirth, disgrace, and some hysterical who knows what else. The contorted bliss-dread look on Jack's face indexes his miserable pleasure. The rest of us leap out of the way. His body shudders: exposure, the private waste of the publicly wasted Jack, rendered public, but with an imperfect indifference.

As if channeling some Rabelaisian (1955) banquet scene or effecting some Sadian (1966) moment of grotesque transgression, maybe enacting Berlant and Warner's (1998) poststructural history of defilement, a lesson in how privacy and shame are mobile and publicly constituted enactments, Half-Jack soils himself. Surprised, he smiles. But this isn't an intentional action. There is no protest or apology. He's not trying to be political in any way, or polite. Rather, Half-Jack is in a state of inarticulate, nonsensical becoming, animal-monster-defilement disorganization. Here the corporeal exuberance of this grotesque body is a changing transforming body, insistently and unabashedly material. Half-Jack is erupting out of himself, overflowing his own limits and exploding his boundaries, in the process of opening connections with sea, sand, humidity, the flamboyant, the ripple sounds of the waves, sharing as much of a recalcitrant force of life, not quite human, mineral, or vegetable, what Deleuze calls "A Life" (2001: 27–30). This is a body that is simultaneously intense smell, grimy sweat, soiled irruptions, accumulating chaos and debris, a space-time of dissolution and transition. While he is not really conscious of his actions, his sudden impulse is still performed as a routine act, but with a difference. This time it is a public performance of a private act. But the power of the performance is in Jack's wild assemblage with any number of disjointed things (trees, sand, heat, humidity, driftwood, boats, water, rum, shit, soiled clothing, sweat) and moments of tourist seduction when in touch with local realities (Creole sociality and life), the result of its own unique combination of connective and disjunctive forces: a conjunctive synthesis of contingent transitions or open co-dependencies of difference in the act of fashioning a shared deformational force field of contact.

No half measures for Half-Jack. His smell prompts an emergent atmospherics that feeds the foul air sticking to our overheated bodies. Smell on air begins to collectively inform us, it circulates around the group, across the beach, floating on the sultry humidity, the rum, the ganja, and into our lungs and muscles, settling in and over things as a collection of forces, sensations, and materialities that becomes intensely, contingently, and nervously generative of something. This is Richie's milieu, he has seen it all before, and he smiles as Jack's smell potentializes some perverse version of the good life that entangles Half-Jack's usually organized world of individual capitalist zeal and touristic good intentions into something different, something that is yet to be, something incommensurate with politics as usual, an incipient body politic that is conjured through the emergent field of Richie contact—some possible useful enchantment, or not—a compositional smell-sensation making new convergences and enlivening what is organizing as the social, right here and right now.

Half-Jack continues his little beach dance. But his smell-power far exceeds his physical control. And it is much more than an act of making the private public as some reflexive play of purity and danger. This act takes an intangible, sensory form as some raw combination of awe, repulsion, pleasure, disgrace. His smell is literally an incipient feeling in the air that catches us up in the moment. We collectively inhale and we populate the field of contact with an intangible excess of sensations that cluster around giddy revulsion. The corporeal scent excess evaporates as we cross a threshold of some supercharged sensation-intensity seducing us into some state of "who knows what." We all feel it and are both repelled and fascinated by the visceral intensity of Half-Jack's ungoverned, undisciplined, monstrous animal-becoming, a body without organs (Deleuze and Guattari 1986a: 160–61). It's a shocking but seductively inclusive moment that slides back and forth from pleasure to disgust, to laughing spectacle, to an absurd putrefying situation in the act of its realization, under Richie's grove of beachside trees. Jack's animal materialization has the group mobilizing a fusion of potential sensory relations of movement that maps a mutant trajectory never before traveled by any of us. Jack made something happen, something outside the certainty of the normal, something ". . . instead on a singular path of freakish becoming leading over undreamed-of . . . horizons" (Massumi 1992: 95). What the next dangerous act might create of life/animal/material/tourist/local co-constitutions and entanglements of a "monstrous becoming-other" mutation shakes us all up: Paradise, unruly and unbound. Our bodies tingle in the moonlight encounter with Half-Jack yet shake with free-floating anticipation. What might these bodies do if left unchecked? And this bears directly on bodily affect—a body's capacity to affect and be affected, to act unleashed, to invent "new trajectories . . . unheard-of futures and possible bodies such as have never been seen before" (Massumi 1992: 101). An unfolding maladaptation of potentials, this is where bodies can go in the extreme, their range of affects, their singular difference, or what Deleuze and Guattari (see Massumi 1992: 98) call their "latitude." Half-Jack is his own anti-humanist wrecking ball of smell, shit, and sweat, a nomadic subject unfaithfully allied to a master tourist trope of tropicalization: an intemperate act of disobedience, an act of betrayal to resolute structures.

We all take a step back. Someone tries to revive Janet. No luck. Someone else tries to direct Half-Jack to the water. Again, no luck. Half-Jack is a seductive monstrous unruly force of dancing becoming, until he collapses in the beach sand, a smelly, soiled mess. Richie, chuckling nervously now while looking a bit concerned with what he has spawned, puts a tarp over the cataleptic Half-Jack and insists that we keep the party going, but Half-Jack's smell rubs up against what's left of the sweet allure of the evening leaving each of us oddly buzzing, in shock, concerned, ashamed, and trying to figure out our own best routes home wondering what just happened or the quickest path to another party that seems to be in full swing down the road, leaving Half-Jack and Janet to their restless dreams. On the way out of Richie's place we speculate about what Half-Jack and Janet will say the next time we see them. Some of the group have been just where Half-Jack and Janet are right now grateful that it isn't their turn again. But there hasn't been a next time, not yet. They quickly and quietly disappeared soon after the party and, apparently, were not seen in Wallaceville until they decided to sell the condo. Everyone is waiting for their return while their condo now is sublet year after year.

Seduction: "Con-tact"

The Happy Hour Hank and Half-Jack stories might be unique forces in themselves if there were not so many others much like them in Wallaceville. Tourists end up very drunk, or sick, or stoned, or naked, or all of that. They get into trouble of all kinds, find themselves short on cash, passed out on the beach, embarrassed, or maybe just a bit roughed up, or regain consciousness and panic because they don't know where they are or who they are with. But that's how tourists find *local wisdom*, Richie laughs sadly, always the sage, *it's how they get rid of their baggage*. The tourists wake up in pain, feeling humiliated, embarrassed, maybe smelling like the anxious excess of a drunken splurge, a lot worse for wear, feeling the after-effects of an intoxicating seduction, wondering what happened, a sense of high anxiety mixed with the relief that they are still in one piece, but ambivalent about their opportunity to share an intimate, if intemperate, moment with the locals like Richie as they beat a hasty retreat out of town, as Paradise collapses into panic. A shared moment of wisdom, all around, that's the way Richie sees it. *The tourists are my research experiment*, he tells me. *We should write a book.*

Such tourist seduction starts as exciting and almost irresistible Paradise exuberance, a lure, the other side of which, when activated, is an ambivalent force of shared banalities, or of violence, or of lower material bodily exuberance and embarrassment, and any of it can be the basis of an emergent sociality or a field of exhausting enervation, or simply just something else to do. Seduction as contact can express itself in any way: trap, decoy, enticement, temptation, attraction, you never know, but as "a moment of wisdom," as Richie would put it, it is a moment in the making, a becoming, and emergent, incommensurate composing. Such moments of contact are felt as pleasures and warnings, as repetitious daily routine or as dangerous and excessive moments of a reckless abandonment of restraint, open-ended and contingent. They may feel like a slip-up, a crash, a warning, another chance, a headache, a rush. For Half-Jack and Janet it's about tropicalized enticements that dangerously shape shift into over the top intoxicating encounters in a new holiday space that they have paid dearly for, in one way or another. Such tropicalizations carry a sensory charge into possible frayed moments of out of control, wild abandon that exceeds the work of Caribbean tourist representations to become moments that matter, as forces gather to a point of impact and instantiate something crazy (Manning 2007: 134–61), for good or for bad. For Half-Jack it is the desire for new connections and the possibility of better local recognition as a kind of reconnection with strands of tropical images that Richie is keen to potentialize as a lively creative capacity. Half-Jack is a demonstration of how hard it is to keep a body from taking leave of itself. Richie's trajectory is to carve out life chances from forces too strong, too suffocating, and too strange to bare. In this milieu of uncertainty, navigating entanglements with the power and control of a tourist state, Richie becomes inventive. His inventiveness is a line of flight, a desire, an escape, a confrontation, with an otherwise local entrenched, blocked and territorialized Wallaceville.

For Hank and Richie it's an odd ordinary that throws itself together out of local needs for warding off boredom and distraction, but more importantly for expressing some sign of a momentary resolute real when tradition and local life no longer carry the charge they once did. It's the seductions of shape-shifting solutions of a sort that help shake the drudgery of the everyday with something, anything, else in a place that is transforming into a monstrous tourist pleasure space unlike anything any local could ever have imagined. It's a Wallaceville where disparate events and sensations come together to form that odd and furtive ordinary, the repetition of which leaves a residue like a habit—a living tropical cliché and another moment in which life becomes impossible. And that's the atmospheric pressure of an emergent present and a form of life that Richie rides like a wave and that opens onto some unpredictable futurity: contingent, incommensurate, uncertain, and conditional. It is when the signs of some real social life give way to some kind of engaged worlding such that new intensities and potentialities fairly shimmer precariously.

What I am doing here is tracking the affective intensity of Richie as a seductive force of "con-tact," a bad example, an emergent vitality that agitates, conjures, and fascinates, that expresses what "tact" activates while "coning" it, a moving target building something insensible out of wild, riotous sensations that incite life as potential forces coming into play in this emergent state of phantasmagorical emergency taking shape as neoliberal exception, on the nervous beachside in Belize.

Note

This chapter was originally published in 2014 as "Mr. Richie and the Tourists" in *Emotion Space and Society* 12: 92–100. It has been substantially revised.

Chapter 4
Nowhere Paradise

In a Senseless State of Pleasure

You may never have been to Wallaceville but if you have seen any of the extraordinary pictures used to sell Caribbean tourism you have a tempting and seductive image of the place. You already know it. Beautiful bright sandy beaches, a translucent, azure blue sea, blue sky and sunshine, you can almost feel the soft warm trade winds blowing across your body: it's relaxing and dreamy. You can swim, sail, scuba, and snorkel or spend your "downtime" in a hammock under the shade of a palm tree sipping beach drinks and taking in the splendor of nature and local culture.

The Belize Tourism Board advertising brochure says that the place was once a "sleepy fishing village" that is now also "an exciting tourist destination, where change has taken place without losing the original Belizean culture. . . . A sidewalk meanders through the village along which you will find a variety of gift shops, beachside bars and restaurants specializing in everything from local Belizean dishes to more exotic Caribbean cuisine. . . . Once you have rested you can enjoy the casual nightlife the village has to offer" (*Destination Belize* 2003). There you have it: a warm, sun drenched day and glorious evening in Wallaceville. It's smooth and mellow and intoxicating, it's what you have always wanted in getaway pleasure, in a hideaway place advertised across the globe as "mother nature's best kept secret."

Tourist Bodies

But it's all more than that, too. Wallaceville is a place that tourists have been getting "high" on for a long time. The stories of hippies "discovering" Wallaceville in the 1960s are legion, but its discovery also has a longer history that takes us back in time to the era of British pirates and buccaneers. Today, all of these stories of discovery act like sediments of romantic discourse that establish compressed layers of cultural images that fuse time and space into fossils that preserve a radioactive quality of original contact to be tapped into in the contemporary tourist moment of encounter. Wallaceville, awash in romantic images, is a place where the body can't help itself. It has a reputation for good stories, food, culture, and nature as well as ganja, cocaine, "Rasta sex," and booze, all tied to a complicated beach party ethos, but in a relaxed, sunny, and laidback mode that sets the rhythms of escape and relaxation. It's a place to relax and let yourself go, which is the only instruction for tourists. It's a place where the body surges, drifts and dreams, gets sidetracked, indulges, falls down, crawls, gets up, slurs, indulges some more, hits the wall, regroups to do it all again, or beats a retreat and gets out. This is a place where the tourist body knows itself in states of intensity, vitality, exhaustion and renewal.

Take the "Royal Windsors," named as such for their conspicuous sense of entitlement and privilege that local bar owners, who gave them the name, consider hilarious. They are a natty thirty-something, highly ambitious, professional couple from Windsor, Ontario who came to Belize as tourists on an impulse but are now caught up in the inextricable forces of the tourist industry, working for some "south seas" global corporate king from Singapore who owns a flotilla of dive boats that he strategically moors at prime dive sites around the world. The Windsors are pumped for the new live-aboard party action their boat will promote in the Caribbean, but especially along the barrier reef of Belize. They have a business plan for sea-land adventure and some local support for their operation and spend their time drunk and stoned in the bars talking up the great fun of their new lifestyle. She gets excited and takes off down the beach with a couple of local "Rasta" guys, shedding what few clothes she is wearing. She likes to feed her urges and loves to indulge. She tells me that this really is freedom and that her work is her play and she lives her life "outside the box." He moves to another bar and gets drunker yet, and to another, and another, until he can't stand up or can't stand the situation any longer and he passes into oblivion, sometimes under the care of a local Creole girlfriend who more or less looks after him now, when she is around and up for it, or needs the money or a place to stay. She is a favorite of the Windsors and always welcome.

Such tourist bodies spin madly out of control daily, beyond any boundary a limit could imagine, in a state of senseless pleasure. The Windsors are in the game as entrepreneur-consumers straddling the line where their private intensities appear like tropical romance images on a public stage and where public representations like "lifestyle," "happiness," "escape," and "adventure" are made sensate with an intimate force that they expend imaginatively for themselves, and now for tourists, as the lure of seaside tropical Paradise. They have bought into and drift on and surf states of all-encompassing pleasures and forays into the world of emergent local and global forces and vitalities. The Windsors are retooled bodies drenched in desire, looking for impacts. They could usually be found at a local bar and restaurant each morning at around ten for their breakfast, a combination of vodka and eggs, madly shovelling matter into force, laughing off the night before and getting in the mood for the daily business of the next rush. It's empire building into their bodies, organizing organs into monstrous bodies of excess. Such is the process of embodiment as it is distributed across a geography of affects.

Wallaceville is a place that makes your skin shiver and tingle and your body flush with the sun, sea, and inexplicable urges and desires. Plateaus of invented life form around the body's dreamy surges that materialize in the rush of tourist consumption. Like one big pressure point, tourist beach bodies feel the shuddering effects of this spectacle of Paradise, taking it all in and matching it up for inner emotional weight: deliriously happy yet a little afraid of how much fun it is, definitely titillated by the array of exotic pleasures wrapped in layers of tourist spectacle, narrative, and imagery. Yet, the dreamworld of Paradise can grow delirious by its own excesses, a concurrent starvation and deluge of the senses (what Buck-Morss [1992] calls the anaesthectic effect of the phantasmagoria), postponing the feeling of dread and loss through a flourish of tourist commodities. The commodity, as a presence-effect feeding the simulated phantasmagoria of a tourist Paradise, "jump starts" life sufficiently to keep it going, to hold its attention, to maintain its course but not enough to propel it to awaken, to reclaim life from another scene of "survivor" or survival. This is what Giorgio Agamben (1998: 155) calls "bare life," an exacting force of twenty-first century biopolitics to make bodies survive, to keep them in the loop of spectacle in order to keep things going.

There is no reassuring middle ground in Paradise. One moment it's more than adventure, more than just passion, more than just fun in the sun, more than just a rush, more than just pleasure, ecstasy, or a holiday. But at other moments it may feel like just another tourist experience. After a while you get a little bored with the humming banality of it all. How many rum punches can you drink, lines of coke can you snort, sex partners can you have, snorkeling tours can you take, ruins can you look at, monkeys can you see before it all starts to feel like just another tourist moment? How do you know when you've had enough or when you have gone too far? At such moments you begin to feel the dampening, empty aesthetics of the Paradise spectacle of adventure and relaxation. Attempting to relate to everything you find it impossible to relate to anything and reach for another intoxicating hit of advertised adventure experience. As Debord (1995) argues, the society of the spectacle controls life through reducing the subject to the present tense in which desire plays off against disaster, with spectacle consumption playing the narcotic erasing the fear that haunts the system enough for life to go on. This world of excess makes the tourist body convulse and spin out of control as it pitches its way through a wild zone of contradictory and ungraspable traces and lingerings, of things happening without any real certainty of what's going on. This is the world of tourist-affect that surges through the body like a shock, a perception an instant before it is finally coded and processed into language as a feeling with a meaning and a structure: a state of blank pleasure that feeds the empire of the senseless.

The key to understanding the ways in which the society of the spectacle both ties the body to power and yet never quite captures it, is affect. The trace of affect is located in the mutually implicating reverberations that catch up the body in the effects of spectacle. It is to take into account the way in which the machinery of human perception and feelings connects the body to the world around it producing an intimate entanglement of the body and its setting (Buck-Morss 1992: 12). Affect, then, is the current that runs through the body like a shock, manifesting itself in a shiver, a tingling, an ache, breathlessness, or a flash of unexplained desire. It is perception before it is processed into and broken down into identified feelings of happiness, dread, pleasure, sadness, and anger. Affect is a productive intensity. To narrate the body through a vocabulary of affect is to realize that the body cannot be fully reduced and explained by qualitative or quantitative means. As Massumi (2002a: 25–26) says: "the relationship between levels of intensity and qualification is not one of conformity or correspondence, but of resonance and interference, amplification, or dampening. . . . Intensity would seem to be associated with non-linear processes: resonance and feedback that momentarily suspend the linear progress of the narrative present from past to future. Intensity . . . is a state of suspense, potentially disruption."

Twitch's Mysterious Death

Then there was Twitch's shocking death. Joseph asked the question that was most on everyone's mind: *What in the world did Twitch ever do that he deserved to die?* Twitch lived all of his sixty-four years in Wallaceville and got his name partly from a case of the nerves he seemed to have always had that was associated with the fact that he lost most of one hand and much of the muscle strength in his extremities in a machete fight years ago that left his arms and legs weak, shaky, and deformed, but which became indistinguishable from his later convulsive alcoholic condition. Still he was able to handle his prized wheelbarrow with little trouble. He would do odd jobs for people, carrying things in his wheelbarrow mostly, like groceries, boat supplies, school kids, or luggage, up and down the main road. That's how Twitch survived. He always seemed to be around when you needed him, always coming and going. Twitch never asked for money but mostly wouldn't turn it down if it came from a tourist. Usually all he wanted in return was a cold beer, maybe some rum, or food, or a place to sleep in a house hammock, or even just a visit. Twitch knew everything in the village because Twitch talked to everyone in the village during his daily rounds. And everyone talked to Twitch. Villagers always went to Twitch for the local news, gossip, and opinions. Twitch used to say that it was important always to help each other out, that community was everything.

Twitch was run over a few years back. They found his crushed body on the main village road in front of one of the popular bars. A very drunk white woman who lives in the village full time ran over Twitch's body. But nobody thinks she killed him. Twitch usually slept down by the docks somewhere, often on the floor of a friend's boat, wrapped in a blanket. They say his body just appeared out of nowhere lying in the road when the tourist woman ran it over. But everyone says it was very strange for Twitch to be in the middle of the main road at all, and there so late at night. Richie and Mr. Normal were not alone in saying that Twitch's killing was no accident. He says that when they picked Twitch up off the street the next day, they also collected two bullet casings out of the fly-infested mess of impacted body parts and dried up blood. Others say they saw them too. These were the incongruous and puzzling signs that villagers talked about over the several weeks after the event.

Someone covered Twitch's body with a boat tarp. The police finally arrived the next day from Belize City with some specialist to pick up the body, take it away for inspection, and complete all the paperwork on the "accident." As in every sector of law enforcement in Belize, the nation has strong anti-drug legislation on the books but can't "afford" (a loaded term, many say) to enforce the laws. Dealers and users in Wallaceville do pretty much what they want with little interference. But so does everybody else. It all serves to support a local Belizean ethos of non-interference in the affairs of others, expecting villagers, tourists, and the new class of beachfront baby boomer property owners and entrepreneurs to figure things out in their own way with only selected state and local help (Sutherland 1998: 45–56). It all creates a "wild zone of power" (Buck-Morss 2000: 3).

It's from that wild zone that locals scanned the village for an explanation. Again, Twitch was rarely seen on that road at night. Some said (but it was never officially confirmed) that Twitch was shot twice in the head and his body dropped in the street to be run over as a cover-up. But why? Miss Grace nervously suggested that, *Twitch probably knew too much about things like drugs.* Around the same time as Twitch's death the village was close to panic and in a buzz talking about the two bales of cocaine that someone somewhere said were missing from a big cocaine shipment off of Gladden Caye. Sadly, Miss Jane summed up village opinion. *What Twitch knew and din't is probably what got him kill't. He was always talkin' around and makin' deals. Me? I'm nervous about what's happenin' here now. None of it's good.*

It's how local villagers (Creole and long-stay expats) in Wallaceville notice everything in a style of a nervous "not watching." Locals scan appearances for signs of "God knows what" from behind shaded slat windows, with furtive, cut eye looks from under hat brims, or from house stoops and hammocks. Like when the brand-new red Mercedes diesel SUV rolled into town a few days before Twitch's death and two guys from *who knows where* started hanging out in the village spending lots of money and time with Elvis, one of the local drug lords. Anthony calls Elvis a *nasty piece of business that no one messes with*. Others certainly agreed. But he also doubles as a prominent property owner, estate and condo developer, a real estate agent, and a trusted acquaintance of several politicians in government. Outsiders who don't really know him trust him until their business dealings with him inevitably go sour. They end up with stiff legal bills and lingering doubts about life in Paradise.

Elvis lived in a house guarded by two vicious dogs, the only one of its kind at the time, in a new village subdivision of tourist beach houses that he was developing. They say he has protection, that he hires hit men. A friend of mine's friend's brother says the guy has paid him for fifteen hits. *It's almost his living*, my friend says with a twisted laugh. Everyone knows that he's a US fugitive, an indicted felon now with Belizean citizenship. *He can't go home*, Glen says. They say that pretty much any expat land developer who has lived in Belize for any length of time is a "pirate," dealing in drugs, land, and development schemes. It's all part of the same assemblage, conjuring the same effects.

Or like the fact that around the time of Twitch's death villagers couldn't help but notice and feel nervous and disappointed about the stumbling bodies of relatives and friends scattered about the village, signs that the crack houses were back doing a brisk business. Also, you couldn't help notice how the coke dealers were back in the bars and restaurants talking up private parties out on the cayes with tourists and buying drinks and dinners for tourist girls like Heather, a thirty-something British woman who is into resource management at home and sex and sun adventures in Paradise.

The tangible objects of this local talk of Twitch's death, then, are the emergent vitalities that appear on the supercharged border between village and tourist life that turn the very border itself into a wild zone of contingent, nervous contagion and flux. My point is that the talk has no meaning in itself. The talk doesn't add up to anything. Rather, it turns into more stories, conjecture, and observations of strange things going on, of weird circumstances, of very odd moments, as these things cross each other, merge, fragment, and recombine into images and actions that address the weight of immanent possibilities culled from mysterious actions, big meanings and the sudden, forceful realization that there may be some inescapable relatedness to things: the force of some new line of flight. Twitch's death is not interpreted, but rather assimilated, consummated, used—or not. Such talk cruises the surface of things as a floating influence travelling through public routes of circulation that have impacts and make a difference that is felt. But this is a difference that is far more fundamental to cultural life and far more fluid than our models of positioned subjects and citizenship have been able to suggest. The difference this talk makes animates cultural forces at the point of their emergence where they charge things up and the nervous system becomes more nervous yet.

Adding up the signs and fashioning stories of the senselessness of Twitch's death spiralled out of control like a wave from one central arresting image: Twitch's mangled body in the road, his blood running out of the cracks in the road to form a dried-up puddle, and the shell casings. It is an image that acted as a graphic sign of the sheer impact of events on lives caught up in a contradictory space of ecstasy and trauma in an encysted local world groping along in the midst of a minefield of global forces of extreme consumption, tracking the traces of impacts on "a nice little place" that has "lost its senses" and "gone mad" for unregulated tourism development in a country that is rapidly becoming a tourist state. Twitch's mangled body is an arresting image that doesn't so much gather force as a type or under some sign of meaning. Rather, it is an acute case that flirts along the surface of things gathering texture and density as it mixes in as this new ordinary life. It is in the space of such an encounter as this that people try get a feel for things going on and make sense of them not by constructing a rational explanation of what happened, but by feeling their way through it and offering accounts of its impacts, traces, and signs.

Just Relax

In yet another frame that shocks the growing and nervous exhilaration of local, consumer, and tourist tropicalization in Wallaceville, new expat neighbors build their gated beach condos so that there is no public space or comportment left. Cozy holiday and retirement cocooning is where forms of Paradise living have become tactile, and the bubble of a Paradise fantasy life born of commoditized "local Caribbean culture" grows sensuously vibrant in the circulating impacts of a tropical dreamworld haven; still another arresting image. And it all places heavy demands on an image. The holiday-retirement house in Paradise guards against the "outside world" with its wild scenes of crime, chaos, disease, and decay. It's good to escape to this hideaway haven to revitalize and find renewal and a new purpose in life. Life's little pleasures hooked up to the big picture of Paradise that settles into a dreamy connection with things, a resurgent image-affect of tourist pleasure mixed with a pension-retirement master plan.

Vitality and happiness unfold as if naturally and effortlessly, but anxiety and fear are the grounds over which it flows when the big picture dreamworld implodes under the weight of its own embodiment and plays itself out to the point of nervous exhaustion. A dreamworld beyond the pale becomes a nightmare (Stewart 2007). That's when the dream is confronted with its own excessive strivings for pleasure, and you begin to ask yourself why you are trying so hard to relax. You cultivate the art of "loosening up." Go Rasta, buck the rules, live beyond the normal limits and restrictions, your mantra becomes "no shoes, no shirt, no problem."

Yet when something horrible happens, like Twitch's grisly death, life in Paradise ruptures and tourist bodies (especially expat bodies) become immersed in a flow of fear and a stiffening case of the nerves. They start to grow tense, blocked, and begin to show shock and exhaustion. That's when the panic sets in. That's when the collective deep breathing begins. Your beach hideaway in Paradise leads to a healthier lifestyle and a chorus of new therapeutic routines that now set the controls for beachfront happiness. And that's when the village acupuncturist is all of a sudden busy, but then again, he's busy almost all of the time anyway, clearing nervous blockages to better life and making the tourist/expat experience super-special.

This all occurs when public specters grow intimate, as the dreams of a new responsible community voice and a reasonable public sensibility are frustrated by corruption, conspiracy, and lunatic flourishes of viciousness marking new threats. That's when there is a rage for new rules, better management, security, protection, government, food security, pet shelter. Ordinary life lingers in the uncertain, haunting realms of Paradise promised and Paradise lost: pleasure mixed with dread.

Once again, the baby boomers are forced to actually think about Paradise, beyond an imaginary stimulus. But instead, they keep busy working compulsively on securing their pets, properties, boats, mental health, and houses. Or they pack up and get lost. Or they anxiously scan the village for more signs of decay and fight for more reasonable instruments of control to dig it out. Multiple lines of flight, different trajectories: follow one, shift to one of the others, veer off yet again. It's a matter of nerve and impacts and how to value a dreamworld investment holding off any shadow of nightmare and disaster.

It's best to keep busy, so they put more effort into stabilizing the Paradise dream. They try to get involved in the community, to help build a cleaner village and propose plans to deal with the vermin, the stray dogs, and the garbage. That's when they begin to demand more and better private security systems. They build their beach dream homes; maybe it's a time-share, with sophisticated surveillance technology, this time guaranteed to keep the local thieves out. They share information about community events and life at the next yoga class or massage at the spa. They play horseshoes with the locals on Saturday afternoons on the beach, making connections with them and feeling good about positive contact; but it's a constant reminder about finding themselves surrounded by a nervous otherness that is never quite assimilated or domesticated, no matter how strong their dreamworld image is of the pleasures of Paradise.

But their involvement means they will never be able to escape the cruel world of corruption and fear just outside their gated or protected homes. They can't buy land or build without some Belizean ripping them off or stealing them blind. No local honors a commitment. Contracts are useless. The law doesn't work. Finally, all sorts of fears—drugs, thieves, corruption, chaos, the sun, the food, the strange weather—and unseen dangers swell and flex their muscles. And there they are, right back where they started, nervously weighing their lives, as the future grows tense and tactile with an unanchored anxiety. It's at that moment that their attachment to this bit of Paradise slips. No one told them about the downside, that it could be so tense or even downright dangerous. Their dreamworld image meets catastrophe with an impact that almost sends them nervously packing. Maybe they want out but can't get out. Since 2008 the housing market for getaway second homes has almost dried up. They can't get anyone to look after their places so that they can get away for a few weeks knowing that if they leave them unattended that they will surely be robbed. And now they feel caught and ask themselves, *now what?*

Tourist sensations work the thresholds of the local real and the image of some generic Caribbean dreamworld of pleasure and adventure, and now the influx of baby boomer tourists who are seduced by the tropics into mad and grand consumption. Seduced by Paradise, and after buying up local family land at outrageous prices, then building their huge, expensive houses, and then becoming unhappy and frightened with the way things are always spinning out of control and with how they always get cheated or robbed by the locals, they get their own case of the nerves, Paradise nerves. That's when they begin to demand more and better security services, community mindedness, garbage pick-up, waste management, dog control, water, neighbors, houses, gardens, local comportment, wine, food, stores, a dress code, roads, and the like. They want to become useful citizens in their adopted country. They are busy dreaming about the day a Target will be built with more convenient parking and when better overall infrastructure might finally arrive.

And what floats along in the wake of this surge of urges, like the garbage the new, super cruise ships jettison along Belize's barrier reef system, are the new burgeoning tourist industries that support such desires for Paradise. They grow exponentially along with a new local panic and fear that things are flying out of control.

Disjunctive Convergences in Belizean Tourist Encounters

Energized by the seductive and calming power of holiday dreamworlds, and sometimes the possibility of a future life in Paradise, tourists visit Belize. While floating on the surface tension of a supercharged ordinary, local Belizeans are learning to "refashion their futures" (Scott 1999 but see Piot 2010). These social dynamics mark the space of tourist encounters, the merging of holiday lives and every day, local lives in Wallaceville. I have described a transitional instance of tourist encounter, which developed through an official tourist discourse that highlights Belize as an international place of world class interest for those looking for nature, adventure, and a bit of Paradise. Lounging on the beach in Wallaceville, signs, capital, and new sensoriums of public culture surge through circuits of cultural exchange seducing holiday makers, expats, and locals alike with an incipient vitality creating a performative space that opens on to new fantasies of becoming, new potentialities.

Always just beyond the control of representation, the stories of Twitch's death are acts that animate cultural forces at the point of their affective emergence. Yet again, they are stories that add *to* even if they never really add up, and they create ambivalence and a jumpy, agitating nervousness central to the encounter between Wallacevillians and tourists. In the face of unfathomable events, like the new local forms of culture emerging around tourism development in Wallaceville, and of unforgettable impacts, like Twitch's death, local life bounces from one piece of wildness to another. Stories collect, creating arresting images that don't gather force into a meaningful event so much as they remain excessive cases gaining intensity as they become part of what goes for the ordinary in Paradise these days. This ordinary is an ongoing moment of enfolding, an unfinished rush toward new Belizean realizations, a place where meaning is in the process of collapsed or capture and tourists and locals alike are left with an active world gesturing at immanent possibilities. Twitch's death is a point of impact that instantiates an event that literally "makes sense" of conflicting and converging cultural forces at the point of their affective emergence.

Life in Paradise is also part of the story of this Belizean present and this is also what I mean when I say that a place like Wallaceville is caught in an ongoing and fabulated present. Such a present begins when this assemblage of things, like the flows of power, drugs, and money out of unnumbered back accounts, the rogue dreams of indicted schemers, culture and tourist industries, NGOs, the IMF, disciplines and institutions, diasporas, expats, crooked politicians, dreamworld image productions, and the like, start to enunciate advanced consumer capitalism, neoliberalism, and global empire and, simultaneously, starts to unravel into the maddening, giddy and traumatic space of a senseless tourist state of pleasure, the pulse of an emergent force at the point at which it instantiates life and things: virtual Paradise.

Twitch's death is a singularity, just as tourists beating a path to Wallaceville's beach Paradise is also a singularity. These are all examples of tourist encounter, something active, some form of becoming. "An example is neither general (as is a system of concepts) nor particular (as is the material to which a system is applied)" (Massumi 2002a: 17). Rather, examples become disjunctive convergences that express a relationship between power and desire felt and observed in the ability of bodies to "become other." In doing so they become bad examples.

I have tracked the generative immanence lodged in a singular scene of a tourist encounter as a bad example. But rather than look for an explanation for things presumed to be parts of some system (world, economic, gender, national, local, ecological, political, or symbolic) or develop a critique of tourist representations, I have tried to sidestep, sometimes drift and flow with, and track intransigent vitalities in bodily tensions, collective excitations, dreamworld absorptions, free-floating enthrallments, in fact, any mode of intensification that emerges as a tourist encounter.

This analysis of encounters is obsessed with states of emergence in moments of nervous excitation, exhaustion, shock, ambivalence, and impasse, and it tracks lines of force as they materialize in such moments, or as they appear in encounters that literally energize tourists and locals alike. It is encounters in the making as emergent vitalities that interest me. I am trying to evoke the moment when an image touches matter to become a moment of impacts. To do so, I follow growing states of excitement, lines of association, the impacts of things, and their effects on other things and people.

This is what Massumi (2002a: 13) calls affirmative augmentation. I am not standing in as your trusted guide to tourist encounters in Belize. Augmentation does not search for answers to real world predicaments or analyze thorny theoretical positions. It is not interested in trying to evaluate things in Belize as bad or better, nor is it interested in the meaning of things. Rather, augmentation is drawn to encounters in Paradise where meaning collapses and we must track the endless wild surge of things, their intensities, and the affects that emerge in the course of encountering.

Chapter 5
Belize Ephemera

Modern society [is] animated by new mythic powers located in the tactility of the commodity-image, the task is neither to resist nor admonish the fetish quality of modern culture, but rather to acknowledge, even submit to its fetish-powers, and attempt to channel them in revolutionary directions. Get with it! Get in touch with the fetish.

—Michael Taussig, *The Nervous System*

Flash/Charge

This is the story of a chance re-encounter with an insignificant thing, a beer coaster, picked up at a beach party in Wallaceville, Belize then found months later while working through the pages of my field notes. The coaster, stuck between two blank pages of a notebook, and my re-encounter with it, had an impact; it carried a charge that worked on me, as Marcel Proust (2006: 61) said objects worked on him, ". . . somewhere beyond the reach of the intellect." The sensations the object aroused in me organized as an unstable presence as they conjured a contact in the form of intense memory flashes. Imagine, a beer coaster, a disposable, functional object produced by the Belize Brewing Company to promote Lighthouse beer by way of the seductions of the Belize calendar girls, the image on the coaster, having such an effect? It was a chance souvenir, a stubborn survivor of another time-place that brought its volatile contents to the present, disrupting the easy flow of things remembered and the smooth coherence of the moment. It agitated as I began to recollect moments of its encounter against the logic of my field notes, working at the edges of the unthought, although vividly felt, as if I might slowly build a story in which to think it: a potentiality, an undecidability, a surprise complexity that adds to and augments a worlding.

Belize Ephemera

Figure 5.1. Belikin Beer coaster for Lighthouse Beer. (Photo by the author.)

The vivid sensory intensities of my encounter with a beer coaster were embodied and located where both verbal and visual archives were silent. Touching the beer coaster again, even such a ubiquitous materialization of the Belize everyday, though it is subject to the deracinating flow of late liberal empire and the censoring process of official Belizean history of capital, commerce, labor, and the nation, sparked the power of a chance moment of possibility, both threatening and enabling. Attending to the beckoning luminosity of this fetish object is to feel a growth that narrative lacks: an incipient composition in the act of organizing itself, a sensuous excitement mixed with dread that made the fetish volatile as an incitement to an arresting presence in the act of its becoming something, something bearing forth as a local history of the present conditions of life, as an event, a moment generated as lively cosmopolitics at the margins of empire.

Thus I turn to the affective charge of my beer coaster as ephemera, as an emergent, non-signifying quality of intention, an unqualified intensity, before it is systematized by narrative to become a named feeling in the body, an emotion, a spoken vocabulary item in a codified language (Massumi 2002a). In experience, affect accompanies narrative, but it is not governed by the same determinations. The force and intensity of an affective charge of my beer coaster is not logically connected to its narrative/symbolic content. There is, as Brian Massumi (2002a: 8) explains, "a disconnect of the signifying order from intensity, which constitutes a different order of connection running parallel." But this intensity should not be mistaken for some romantically raw experiential richness formed in the body. Intensities are asocial because they have not been folded into a system that names them as feelings, but they also bear the trace of past actions, including a trace of their contexts organizing in the flesh. Such intensities, the affective qualities of my beer coaster, conjured other imaginings, other effects that set off titillating chains of felt associations not yet fully narrativized, but that still act as resonant sensations, as forces capable of exerting a deforming impulse on cultural and linguistic codifications as tropicalizations.

The exciting danger of my beer coaster as ephemera is in realizing that it could assemble histories of material-semiotic entanglements yet untold. Such entanglements can agitate those histories we know and generate new potentialities that affective intensities make possible, and thereby shake the security of one's place in the world while conjuring "things to happen" out of the intersections of flesh and object (see Haraway 2008: 18–19). Here the materializations of my beer coaster act "as an abiding yet changeful presence, traversing distinctions between human and non-human, organic and inorganic" (McLean 2011: 591; c.f. Barad 2003). This is what Erin Manning (2007: 134) calls "sensing beyond security" or what Lauren Berlant (2011) calls a "cruel optimism," or how an object and its scenes of desire "matter" not just because of their narrative content, or their representational power, but because the thing encountered holds out potential and promise as it becomes a seductive means of keeping clusters of affects attached to it, for better or for worse. That's my beer coaster, and I began to track its affective charges through disparate scenes that gathered as stories, even as such scenes also remained dispersed, floating, re-combining, regardless of the signifying work they did.

The point is to slow the seductive urge to move to representational and identity thinking and evaluative critique long enough to find a manner of approaching the complexity and uncertainty of such stuff as it exerts a pull on us, as it tugs on us, in order to fashion some form of address adequate to its complexity. This means building an idiosyncratic figure of the material-semiotic connections that stuff produces as singularities; not representations, but, as I have said before, "actual sites where forces gather to a point to instantiate something" in the flesh (Stewart 2003b: 1). It means attending to this figure of connectivity as an assembling, in this case a disparate beachfront party scene that pulled the course of writing about a beer coaster into a tangle of trajectories, contradictions, and disjunctures. Such a tangle included the impacts of "arresting images" (Stewart 2003a), such as the day when a mystery boat floated into Wallaceville Bay, inciting a nervous euphoria.

The mystery ship is equally ephemeral, and I track the way it conjured strange feelings of sudden appearances that made otherwise ordinary life in Wallaceville nervous and haunting for locals, expats, and tourists alike. The haunting sensation that this event summoned, and the hopes and dreams of a tourist imaginary that grew as a cluster of impacts, as such sensations became narrativized, are the result of a promise of vital connection as locals, expats, and tourists go with the intoxicating flow of tropical intensities: seductions, attachments, and impulses. Intensities like these, make the social, whether good or bad, out of the ephemeral qualities of these arresting images and the incipient stories they inspire.

Anthropology's recent efforts to engage tourist imaginaries as a useful concept have shown a powerful predilection for recognizing their socially created representational qualities and their deployment in the act of cultural meaning making, in the way that Leite (2014: 261) explains Bruner's thinking on the function of (meta) narratives. Imaginaries act to influence collective behavior as underlying interpretive maps and to create and transmit meaning more or less smoothly and in concert with personal imaginings and the global culture industries that act to circulate images of others, natures, cultures, artifacts, histories, adventures, and the like. It is to the significance of tourist imaginaries and their role in the transformations of tourist experience, the local lives of those "toured," and the political and economic effects and configurations of such operations that interests most of those working on such things.

I want to ask another kind of question, one about the emergent qualities of tourist imaginaries, and about the generative forces of such emergence, about tourist imaginaries in their making while holding close to my beer coaster. There is a danger in new approaches to tourist imaginaries taking their cue too exclusively from the socially relational logics of objects, images, dreams, technologies, flows of power and meaning, institutions, and capital that are already encoded into one theoretical or methodological story line or another and so these approaches become calculation and definitional practices, matters of fact-making that adhere to regimes of knowing organized under the "three pillars of anthropological representation" and insight: negation, identity, and being (Ochoa 2007: 479). To do so, I argue, risks subordinating consideration of tourist imaginaries in their incipient, sense uncertain, eccentric moments, as unstable and emergent vitalities, generative and activating forces that are not (yet) coded culturally, socially, or politically. Attending to imaginaries in the making, in this case the affective potentials found in a beer coaster and then in the appearance of a mystery ship and then finally in terms of expat ordinary life in Wallaceville, means getting at the intersections of flesh and object and their lively and entangled couplings and assemblages as incitements. It means tracking intersectional co-shaping potentials that incite as they instantiate and materialize things. Paradoxical and unfinished though these things may seem, lingering in the event of imaginaries in the making means lingering in the infinite event of becoming that seethes with uncodified potentials. Favoring an approach that focuses on assemblages of affective forces rather than the political dialectics of representational strategies (such as political, economic, or cultural strategies and models) more familiar in tourism studies, and the dominant focus of the work in this volume, means sidestepping reason and critical analysis and instead drifting along through singular sites of collective excitation while becoming attuned to their affective atmospheres (Anderson 2009: 79; Stewart 2010a).

This is a methodology that is an augmenting composition of imaginaries of Paradise. But I am trying to get at how these imaginaries throw themselves together in moments that are already present as potentialities—something waiting to happen in incommensurate objects (beer coasters), registers, circulations, and publics (a mystery ship's sudden appearance) or the way an expat ordinary in Wallaceville composes itself as an assemblage of discontinuous yet mapped elements or what Leite (2014: 249) usefully calls the "unpredictable forces that lurk around the edges of 'Paradise.'" Such assemblages can be traced through generative worldings of all sorts, like dreamworlds, ways of relating, impulses, encounters, distractions, and seductions of all kinds (Stewart 2008: 73).

This means thinking about tourist imaginaries in terms of their material affective intensity as "vibrant matter" (after Bennett 2010; see McLean 2011, but c.f. Povinelli 2016), flesh and object attuning and attaching in the act of singular, materializing composition, before they are completely narrativized as cultural representations or as political economic forces. This means that I am indebted to the history of compelling ethnographic work that Salazar and Graburn usefully map out in the introduction to their edited volume *Tourism Imaginaries* (2014). It also means that I sidestep a dialectical approach to imaginaries in the way Di Giovine, in that volume (2014: 147–71) and elsewhere, has developed it as an *imaginaire dialectic* or in the way Swain (2014: 103) develops the concept of *imaginariums*, as the "circulation of personal imaginings and institutional imaginaries," while trying to retain something of Di Giovine's dynamic and processual approach to production and reproduction and Swain's image of how the various strata of identity, history, politics, and storytelling pile up and rub against each other producing new forces of connectivity.

It also means that I take one step back from Salazar's (2012: 864) notion that imaginaries are "socially transmitted representational assemblages that interact with people's personal imaginings and that are used as meaning-making and world shaping devices" while still trying to suggest how an assemblage does not become a socially transmitted cultural representation in the making but becomes one lodged in the flesh as impulse, tension, sensation, connection. I am not interested in how a representational assemblage shapes emotional experience or where it comes from. What interests me instead is thinking about assemblages as discontinuous and incommensurate elements of image and flesh that throw themselves together as an affective intensity, as an eventfulness composed out of connection that expresses itself through a relay of sensations actualizing as a weak ontology of lived collective fictions made up of trajectories, differences, found affinities, and diacritical enjoinings (Stewart 2008: 72–73).

My work here is more in line with this "new materialist" approach to tourist imaginaries. Theoretically and methodologically it is akin to Saldanha's work on tourism and the Goan rave scene. The rave scene in Goa, Saldanha (2007) suggests, is a site of excitation and incitements conjured out of fuzzy assemblages: practices, sounds, substances, dance skills, sunshine, and drugs. He works the complicated and complicating contact zone of culture and phenotype to track and evoke whiteness in Goa. Saldanha is not happy with the prevailing conceptions of racial difference as a social construction, rather he focuses on the "viscosity" of race, on the ways in which flesh and bodies gather together for pleasure and transformation. Here race is a dynamic event arising from a complex field of embodied encounter, the fundamental complexity and contingency of which can never be settled.

Wallaceville, as a growing tourist destination in Belize, is a place of impacts, rapidly transforming before everyone's very eyes, as if by some horrid and seductive magic, into a tourist spectacle of some homogenized, global, dream world adventure, a pirate Paradise, connecting locals with the spasmodic effects and currents of global flows of images, capital, stuff, people, and culture. But I am not interested in debunking, with local objects and events, the several definitive structural features of late liberal global empire and capital and the meaning and effects of such things on local Belizean social life, as they become taken up as the stuff of Caribbean tourist imaginaries. Rather, I want to begin to describe the forces of Caribbean imaginaries (see Feldman 2011; Sheller 2004a and b), or tropicalizations (after Thompson 2006), as part of an "unfolding moment in which countless things are being actively generated as fugitive, shifting, indiscriminate, unsteady, and unfinished" (Stewart 2007: 1–2), in the rush and pitch toward some active realization yet uncaptured by what Deleuze and Guattari (1986a) call "identity thinking."

This is akin to the way in which Benitez-Rojo (1992: 2–3) imagines the Caribbean. He describes it as a space of Chaos, as a discontinuous conjunction of unstable condensations, growths, decays, and half-told colonial and subaltern stories that froth and surge together like a turbulent sea that conjures the eddies and pools of seething clumps of frayed seaweed, garbage, flows of sunken drug boats and pirate stories, flying fish, mysterious strangers, tourists, downpours and hurricanes that combine as networks of awkward engagements. Such Chaos gestures toward things still unrecognized and ungathered as concepts, and so they sidestep, shift, and drift uneasily as singularities with unique textures, densities, and mixtures of intensities that scuttle along the margins of events and experience, which suggest how force becomes sensate as it becomes caught up and into the good life tropicalizations of Paradise, or not (Stewart 2003b: 1–2). It's about how a found beer coaster reignited the affective dynamics of a time-place as a chronotope (after Bakhtin 1981), caught in the flesh, radiating force and new potentialities.

Combining Bakhtin (1981) and Bergson ([1912] 1991), the chronotope can be considered a material assemblage of images with duration that contracts into a spacialization and density, intensified but undifferentiated as a logic of random disjunctions. The challenge here is to dwell in the event of a tourist imaginary in the making, not as a reflection of power relations or inequalities and how these are produced and reproduced dialectically or narratively or how exactly they are troubled, contested and transformed, but as affective intensities that parallel imaginaries as representations as another mode of thought different in nature and constituting a space of potentiality where new forms of life begin to emerge, what Povinelli (2011: 6–11) calls "spaces of otherwise."

Ephemera

Two OED definitions of ephemera are: 1) "of a fever lasting only a day" and 2) "something which has a transitory existence." Tourist contacts and encounters with stuff that includes postcards, menus, advertising brochures and pamphlets, magazines, napkins (serviettes), ticket stubs to events, places, and for travel and, of course, beer coasters, all evoke the sensations of the feverish and the transitory. This stuff is the clutter of travel. It attracts, seduces, impresses instantly; it heats things up and flashes as a sensation with a transient impact, it creates an affective charge as it assembles flashes of feverish desire and potential, before it piles up in the corners of tourist attention, spent, exhausted. It is excessive stuff, the ruins of culture, the stuff that once shouted its presence and function as it flashed by as dream world enticement and wish image, even as it ends up scattered across hotel rooms or is found at the bottom of suitcases or in field notebooks, like fossils with a "radioactive" trace of life conjuring a history of some other time-place (Benjamin 1968; Buck-Morss 1991). The point is to dwell in such stuff, and chance re-encounters with it as found objects of another time-place in order to track its affective force conjuring sensations that broadly circulate and give past events a continual motion of relation, contingency, and emergent potential.

Tracking the life of Belize ephemera means tracking its impacts, not so much as meaningful material culture but, rather, as commodity incantation, emergent forces of copy and contact, or what Walter Benjamin calls "profane illuminations." In his *Arcades Project* (1999), Benjamin insisted on the nomadic tracing of dreams still resonant in discarded material things. His process of writing captions for found fragments and snapshots, gathered into loose assemblages, and the way his thoughts pressed close to commodity objects in order to be affected by them, illuminates the idea about copy and contact.

My beer coaster, for example, immediately brought me into agitated contact with an evening in Paradise when members of The Belize Jewels, a group of twelve calendar boys/strippers, entertained tourists at a beach bar. I want to track the free flow of Belikin Beer coasters around which the shifting impulses of tourist imaginaries of "the good life" in "Paradise" grew vibrant as desire and seductive promise. More generally, tracking tourist ephemera in Belize as arresting, affective forces—the contact zone of dream world enticements rubbing up against the transitory nature of a local world gone dreadfully wild for tourism—means holding my beer coaster close and following its felt impacts and half-realized effects and incipient attachments (see Berlant 2011).

Sweet Sunsets

The Sweet Sunsets Bar and Grill, once a popular lagoon-side watering hole just west of the public docks in Wallaceville, used to pump out "the good life" with an intoxicating sensuousness that was eclipsed only by the excesses of its over-stimulated patrons. On any given evening throughout most of the year, tourist and local bodies moved with the surge of the crowd, floated lazily on the pungent aroma of *gunga*, around the bar and settled into dockside hammocks and loungers, taking in the vibe, swelling in intensity with the Rasta music, the flow of alcohol, the effects of the drugs, and the emergent, potentializing rhythm of the scene. Wallaceville and the Sweet Sunsets are places out of any dream world image of tropical Paradise, places that realize the architecture, textures, and material culture of Caribbean pleasure. Relax. Get with the image, with laidback smiles all round, carried on a warm breeze. It all makes your body tingle as you melt into the groove, dizzy with anticipation. You can feel yourself losing control even while settling in. There is an intense attraction to this edgy ecstasy, a giddiness mixed with a touch of alarm at how good it all feels, and it acts as a contact high that seems tailor made for a tropical night in Paradise. This is a surge of the good life as Paradise, and it fits with the image of Wallaceville that is sold worldwide these days.

Party Boys

Flashback: The Sweet Sunsets was pumping out its version of the good life on a hot and sultry evening in mid-March of 2004, when four calendar boy models from the Belize Jewels performed at the bar. The entertainment was sponsored by a local women's group and appropriately named by Alice, one of the longtime expat owners of a popular beachside bar and grill, "Wild Things of Belize: Lady's Night Out." The Belize Jewels were the Belize Brewing Company's male calendar boys, the male parallel to the Belikin Beer calendar girls, but the men actually strip during performances while the women, on the few occasions that they appear in public, stage bathing suit fashion shows. The event was a locally inspired ploy by the women's group meant to showcase local, young men, and identify those with aspirations and bodies attractive and agile enough to become Belize Jewels for the 2005 calendar; a sort of community women's group work project for male youth that hooked up with Belize's "Dream's Come True" Productions. Qualifications? To look hot enough to be able to fill out the demanding contours of a Melenie Matus designed swim suit while holding an infectious smile and an ability to dance and pose. Matus is a Belizean fashion designer with "a knack for comfortable swimsuits and tanning-wear with sex appeal" who was then making a splash in the global beachwear fashion market.

The show started around nine when the Jewels appeared on short bar tables provocatively stripping down to their swimsuits, showing their well-oiled and muscular bodies, beaded with sweat in the intense heat of the jam-packed bar room patio. This was all a tantalizing preview of what was to come later when the Jewels put on a "for women only" strip show down at The Pelican Bar. When the Jewels arrived at Sweet Sunsets the place went wild. Local and tourist women made riotous moves to get as close as they could to the four beefed-up models, the dreams of "drifting off with one of them," as Alice put it, carrying them along on waves of desire fed by the intensity of a mix of intoxicants, desire, and ear-splitting reggae music. Bodies surged forward toward the boys, limbs flying in every direction, struggling for attention while the Jewels moved their bodies seductively, grinding and smiling broadly to shrieks of delight. They threw calendars and beer coasters to the audience. Women, their bodies building intensity out of layer upon layer of sensory impact, looked both shocked and thrilled to find themselves in the driver's seat, grabbing for the men, the calendars, and the coasters all at once. While from the seats in the back, the men in the audience became the noisy rumble of cat-call hilarity and ridicule, picking up coasters and flinging them back at the men. The coasters flew everywhere.

Things looked like they were about to shift terribly out of control when Chas, a very large and local Creole man and the only self-identified gay person in Wallaceville at the time, appeared from nowhere dressed in drag as a caricature of Aunt Jemima. Wearing a pair of revealing bright pink knickers under her trademark servant-girl dress, along with her checkered head scarf, she was an arresting image, a force of menacing and jocular material contagion and potential. She couldn't have been happier with herself or with her imposing, heavy weight impact on the crowd. Chas's Aunt Jemima conjured a nervous moment when the pleasure of looking was retooled as all at once obtuse, erratic, pleasurable, and shocking and as the excited audience cut back and forth between global images of race and gender and some image of a local world caught awkwardly in their grips. Chas's image was a hilarious monstrosity that hit the senses with a mesmerizing, seductive, and repellent impact, a graphic mode of delirious fascination and laughter mixed with a hint of danger and aversion, a complex force, an assemblage of influences as some emergent vitality, in excess of any ideal representation of an ideological structure and way beyond the obvious meaning of a message.

Aunt Jemima's role as MC was to sustain the excitement, prolong the ecstatic moment of beachfront pleasure. One of her strategies was to tell jokes about her seductive body with punch lines underlined by lifting her skirts and throwing out "sweet kisses" and beer coasters into the audience. Handfuls of beer coasters went flying everywhere. I caught one but didn't think anything of it at the time and absentmindedly stuck it in my pocket.

Chas's Aunt Jemima image is a half-written sign of a discontinuous local social world and its history crossing with the trajectories of a global, imperial one to become one of those sites where forces cluster into a nodal point of impacts that incite and instantiate things, much like being right where you are, only more intensely (Massumi 2002a: 10). It is how the shifts to a tourist-based, late liberal capitalism have encouraged new affects and scenes of desire that are actively becoming tourist imaginaries inspiring new narratives of race, gender, sexuality, and biopower that actively free new libidinal energies, new forces of becoming, and new examples of agency that are far from straightforward, but call on a jump and surge of affects when things resonate as both potential and degrading threat: a powerful force of the dream world good life in the making, contingent even if calculating, actualizing forces into something, some power.

In this contact zone, the body gorges itself on an excess of exotic pleasures and impulses metamorphosed into some monstrous "phantasmagorical enjoyment," outside the law of civilized life, in a state of exception, what I want to call a tourist industry "duty free" zone. This is life living up to the force of seductive tourist advertising, enough to be pulled into the loop of global, imperial forces that morph right along with it. But here, life moves way beyond the advertised helter-skelter, toward some threshold of vanishing, toward a death that will not die, an apocalypse eternally deferred: the good life as bare life, as desirously ephemeral as it is.

The shock and applause for Chas's Aunt Jemima image and performance focused everyone's attention on her as she introduced the local young men auditioning for the Jewels. As they were introduced individually, each performed a seductive dance, ripping their shirts and shorts off, down to their new fashion swimsuits, to the music of Bob Marley. There was loud applause and laughter as family members, friends, and tourists alike screamed encouragement. But there was also a lot of teasing by local male friends, expats, and tourists. These faces read like an ambivalent mixture of embarrassment and delight, alarm and elation, agony and bliss as they grabbed at the coasters and sent them flying like Frisbees at the contestants and across the patio.

Afterward, Jimmy, one of the local guys trying out, could only say *I nevah know it was goin' to be so good man, because it actually came out well and I'm fine doin' it. But I look foolish man, real foolish and nervous. I don't know what to do. I don't like it. But the girls like it, ya man! It was lovely. I feel good about it that way.* Jimmy's moment on the big stage potentializes things to become a trajectory of impulses that affective forces might take if left unrestricted or unhindered: a singularity or a matrix of forces actualizing as something.

Local men like Jimmy have been shocked and seduced into this new imperial edge-world of chronic, late liberal, capitalist blur in Belize, now almost completely fueled by the Belizean tourism industry, and touched by the threat and promise of a tourist culture of spectacle consumption frantically pumping out Caribbean representations as tropicalizations. Jimmy was mixing drinks at the Sweet Sunsets and decided he had "the stuff" enough to audition to become a Belize Jewel. The Sweet Sunsets was built five years before the hurricane, and was rebuilt right after it, by two middle-aged English women who had fallen in love with Wallaceville and the thought of a life in the tropics. Their tropicalization of the place included hiring young, good-looking local Creole men like Jimmy to attract tourist attention, especially women's attention. In fact, the women owners hired and named the bar after another Belizean named Sweets, and they shared their Sweets between the two of them for years. *We have our Sweets every day*, Alice chuckled one night while watching him eagerly mix up drinks and conversation with a group of tourists at the bar. Harriet jokingly calls Sweets her *love toy* and her *sugar*, but on that night in March Jimmy was their *crown jewel of the bar*. Jimmy took an awful ribbing about his decision to audition from many of the local men who would have liked nothing better than to have replaced him, or find their own *sugar mommas* like Alice and Harriet, even as they are seduced and recoil in horror from the image of the sexy, carefree native that goes along with the performance. It's as if locals like Jimmy and Sweets are literally touched by the intimacy of the scene and make themselves its convulsive possession.

Stay in the "tropical, happy-go-lucky, no worries" pleasure frame. Tourists love it. Like many of the locals I spent time with, Jimmy was eager to get out of Wallaceville. Becoming a Belize Jewel was his immediate image of freedom and his strategy of escape, this time into the seductive world of international modeling. *Just like on TV*, Jimmy said. His public audition gestured toward a magical pleasure world beyond Belize; a world of success, high fashion, and money. If he could get the chance to show his body to others, he could show everyone just how handsome and "gifted" he was. Impacts like these make up the plateaus of the social for Jimmy. And, in the space of tourist encounters, on a makeshift stage at a beach bar in Belize, ripping his clothes off while in "serious party mode," Jimmy gets high on an intoxicating pleasure and desire with a confidence that ricochets wildly between the raged extremes of a disquietingly hard life and the thrill of a reckless, seductive dream.

This is what Ann Cvetkovich (2003), in her work on public feelings, calls "traumatic realism." On the one hand, there is abjection. Jimmy sends handfuls of beer coasters into the crowd at the bar while dancing his version of seductive moves. Yet, on the other hand, there is a vital move to take on this force with intensity enough to capture it in the senses and make something of it. This is intensity disconnected from the signifying order. It is part of an assemblage of intoxications. For Jimmy, the impact of such a tropicalization of locals—wild, energized, savagely in rhythm—fails to represent the inflicting force of tourism as an object in the symbolic order of imperially inspired capitalism, so he feels compelled to live on and repeat those images. Yet, there is also Jimmy's confidence to take on the force of such images as a subject rushing to bear an affect toward some affirmative actualization, to make something out of an uncertain sense of circumstances: movement toward a sense of maneuverability and sense maneuverability. But this is no dialectical movement as much as it is an assemblage characterized by shifting configurations of flows and intensities in which things are constituted by a play of forces that take possession of them. In this sense, Jimmy's lack of fullness as a subject, his extreme vulnerability comes not from a disconnection with reality, a slipping signifier, or from false consciousness, but from the promise of some vital connection as he goes with the nervous flow of such intoxicated moments of impact and new attachments (see Berlant 2011).

In a couple of hours the show was over, and the excited crowd floated on a force field of ecstatic energy toward the women's show. An aching hush came over the bar patio. With the music turned down you could begin to hear the surf again and feel a faint sea breeze against the buzz of overstimulated bodies, which faded as the crowd drifted down the beach.

The Mystery Ship

But the ephemeral agitates in other registers too, like in the productive predicaments and augmentations of other arresting images of the tropical that have impacts. The day after the strip show, while walking along Wallaceville's beachfront with Aiden toward our usual afternoon, after work, watering hole, the Drunken Pirate, he stopped to point out an impressive ship at anchor in Wallaceville Bay. It was as if the boat had magically appeared. I said something to Aiden about how strange it was that at one moment the bay was clear of ship traffic but when I looked next there was a big boat suddenly at anchor. I thought it looked like a research vessel. Aiden laughed, *That's the magical mystery ship, at least that's what we call it here, because that boat always seems to appear, like "poof"!* No one, including NGO officials, the village council leader, bar owners, tourist operators, village police officers, or the local head of the Belize Industry Tourism Association, who doubles as Wallaceville's newspaper snoop, seemed to know why the mystery ship made these occasional ports of call, although everyone had a theory.

The silent appearance of the boat was a disquieting event. And while we drank our beers talking about it, I found myself nervously playing with my beer coaster. The familiar image of the calendar girls rubbing up against the mystery ship had an impact. It conjured an unsettling conjuncture of effects, a singularity or what Massumi (2002a: 17–18) calls a "disjunctive self-inclusion: a belonging to itself that is simultaneously an extendability to everything else with which [these things] might be connected." Four days later, *poof*, the mystery ship disappeared just as mysteriously and suddenly as it had appeared. *It's still there you just can't see it,* Julian insisted as he began a dreamy, laidback conspiracy riff on the imponderables of space age technologies he'd heard about from some tourist he knows who works for NASA.

Theories about the boat's appearance and disappearance became an elaborate set of stories. Some of these made more sense than others, but all of them were part of village-wide gossip that transformed into waves of free association as mystery ship stories moved from person to person and up and down the cayes. The boat was a drug boat operated by the Columbian drug cartel waiting on bales of cocaine. It was a National Geographic research (tourist) vessel counting whale sharks. It was a boat measuring channel depths in preparation for a dredging operation for the new mega-cruise ships that have now seduced local Belizean politicians and international tourism entrepreneurs further into the seamier sides of late liberalism and Caribbean capitalism. It was US Homeland Security or the CIA or the FBI. It was a Russian-Cuban spy boat. It was a terrorist boat. It was an Italian millionaire's leisure boat; he sails to Wallaceville every year at this time in March. It was an Australian charter dive boat. Some said they knew of the Australian "skipper," apparently a drunk and a letch, but no one could remember his name and no one was sure they had ever actually seen him around.

The people of Wallaceville, both long-time expats and locals, made sense of the boat through their stories, not by constructing explanations for its appearance and disappearance but by offering accounts of its nervous impacts and of its mysterious traces and effects. And as the stories piled up like shipwrecks on a reef, rocked by waves of telling and retelling, the event of the talk formed a tidal rush of dramatic and excessive images that overwhelmed the merely referential. These stories added a vivid intensity to things even if they never actually added up to any one reasonable explanation. And that is how they continue to spread.

Jimmy thought that the mystery ship might contain a film crew. He had heard that an international crew was supposed to film the strip show, that a film producer was searching for a new film star, or maybe it was a new fashion model. Jimmy was hardly interested in why they were there. More than most of the local men and women of Wallaceville, he desperately desired to "see the world in style" and he felt that finding work on the mystery ship might create for him a whole new international life of leisure, something modeled on his favorite reality TV show, *Survivor*, or some new life that felt like the adulatory exhilaration he experienced during his strip show dance performance. He saw the boat as a sign of international pleasure, a sign of a seductive and magical world way beyond anything life in Belize could offer him. If he could only get on board, he could show everybody how to party "Belizean style" and how sweet life could be with this Creole Lothario. In this frame life could be sweet. The corollary: "locals" out of costume, out of focus, those who slip out of frame of the emergent tropicalizations of the place as Paradise are a threat; another image of impact that I will get to later.

This scene grows tactile and dense as it literally and nervously matters to Jimmy, and he in turn is possessed by it and the emergent cluster of sensations activated and actualized through it. And in the intersectional space of tourist encounters, Jimmy gets high on the lure of an enticing fantasy with an intoxicated confidence that bounces chaotically from the realities of village life to the tropical seductions of a globally circulating tourist imaginary. Here, it is the charge of nervous desires pulling into an alignment and some visceral complicity of Jimmy laying claim to a composed tactility, literally into something that matters. The precariousness of Jimmy's life takes form as precarity pulls and aligns some contingent collection of forces, sensations, sensibilities, and materialities into something tentative and generative, "the transitional immediacy of a real relation . . . coincident, but disjunct" (Massumi 2002a: 5). Here an eventfulness is incited, and you can find its shimmering shapes in these scenes of attachment as things that matter cast an uncertain line along the hopeful edge an emergent imaginary (see Gregg and Seigworth 2010: 2–4) .

But the mystery ship was a contagious image that flashed up uncontained by meaning or fact. Story subjects, objects, and events became performers in a spectacle that exceeded reason and the trajectories of cause and effect or truth and lie. The ship is a singular affective intensity, a force, an unchecked impetus, which then becomes an event that literally "makes sense" of that force as matter-forming: affective and material emergence. The power of the storytelling that focused on the boat was that it drew listeners into a space of tense and lingering impacts and unseen but certainly felt forces, an affecting presence, a cultural poetics in the act of making something of itself as an eventfulness. Mystery ship stories and all of the things associated with them became acts of creative contagion (Massumi 2002a: 19).

As such, the mystery ship served to focus anxieties and wild speculations about tourist encounters and about the dreamy seductions and nightmare chaos and contingencies of the world "out there." This was evident in the excessive exchange of conspiracy stories about US spies, drug dealers, strange tourists, terrorists, crazy locals, and corruption, a flow of narrative that conjured a nervous dread that agitated the smooth, tropical vibe of the place. As such the boat super-charged a troubling state of suspense and suspension that haunted the place and its people, pulling them up short, and that lingered as a troubling impulse struggling to "make sense" and make something of things more generally.

It was the impulse to make something sensible of circumstances and events by fashioning stories about the mystery ship's appearance and disappearance that turned the sight of the mystery ship into a tactile force that entered the local senses, lodging there, growing as an intensity, forming into a state of nervous suspense filled with resonance. The mystery ship figures into the force of what Gilles Deleuze calls affect, the double-sidedness of things where, as Brian Massumi (2002a: 12–18; c.f. Massumi 2003) explains, the virtual meets the actual, and where what "matters," as a materialization of local life in the making, is the permeable edge of potentiality itself.

The mystery ship is a "tactile image." It is felt as a sudden, nervous interruption into the everyday world that then floods the senses with a proliferation of intense feelings, tinglings, images, and stories transforming matter into uncommon sensate affects that constantly exceed the requirements of reason and order. The tactile image assembles things not so much in an orderly or systematic fashion as in a manner that forcefully compels. The mystery ship's arrival and departure conjured an intense reverie mixed with dread that mimicked the dream world and nightmare that is now a constant in the continuous euphoric panic that haunts Wallaceville's move into the big picture of world tourism these days. And then, poof, like a reverie, like a dream, like in a trance and in a daze, it was gone.

And that's when the mystery ship mainlines into the flow of things at the Drunken Pirate, making an impact. At least that's what happened to "Captain Ted." Ted is a retired, mid-sixties, middle management insurance adjustor, expat from Nebraska, distinguished locally by his spurts of outlandish and colorful behavior. He has a nice retirement home, built and converted over the several years he has lived full time in Wallaceville, a boat, property, and some business experience in Belize. He was at the Drunken Pirate the afternoon the mystery ship appeared. That's when he linked the mystery ship with cruise ships. This free association conjured dread and excitement as Ted and the rest of us turned the conversation to cruise ships potentially anchoring in Wallaceville. If that happened, the only mystery left would be the one about who most stands to profit. Cruise ships could open business opportunities, but they could destroy Paradise, but property values will sky rocket or maybe not. But cruise ships would really put the place on the map. Who knows, there might be a major hotel chain ready to buy into the place, develop a new infrastructure, formal and reliable sewage, gas and water services, upgraded telecommunications. But it's all so *up in the air* and *there are such good points to be made pro and con cruise ships*, and *it really is up to the Belizeans, you know. But then again, their leaders are all corrupt and in politics for themselves and for the bribes from foreign business interests or the drug cartels.* Ted's thinking about selling now, maybe moving on to Panama, he has friends there, or back to Nebraska or to Cuba now that it is opening up. But he hears rumors that all the best deals in Cuba are gone. Profit or no profit on his imagined property sale, Ted's tired of all the pressure, graft, and corruption and what it engenders materially, emotionally—a nervous, smoldering random anger and sadness.

Maybe it's best to *lay low*, Ted tells me. Yet he can't help but make the weird connection to the mystery ship when he thinks about his place in Paradise and he rides his own intense ups and downs of a strange happiness mixed with free-floating anxiety, like a pelican rides the air. Stick to conversations about the weather. Don't get involved. Remember the master plan. Don't panic. That's life. Take a deep breath. Have another snooze. Get drunk. Have a toke. But the specters of Paradise keep reappearing to haunt the dream, like something out of the corner of your eye, some blot, something unassimilated, and not quite exhausted by what Lacan (1977) calls the Symbolic, something monstrous, something inhuman but breathing, an abject surge that points to a troubling vulnerability that haunts deep Paradise. It's excessive and virtual, nothing you can put your hands on or completely figure out as, again, things nervously add to, yet never add up, and it makes Ted nervous, angry, fearful, sad, and giddy about his image of a tropical Paradise.

The rumors about big time tourist development proliferated at a panic rate in Wallaceville at the time of the mystery boat. And all of a sudden, the image of Paradise became indistinguishable from the sordid world of the international tourism industry. The point is how the virtual and the real collapse in on each other, their difference no longer relevant in the state of emergency that is rapidly forming as the generalized condition of Paradise, grown on the margins of empire, teasing and encouraging new forms of life and new becomings as an emergent tourism cosmopolitics (see Stengers 2005).

An Emergent Tourism Imaginary

What I am trying to accomplish here is something akin to what Isabelle Stengers (2005: 995) calls cosmopolitics, where the cosmos instantiates itself as "the unknown constituted by multiple and divergent worlds, and to the articulations of which they could eventually be capable, as opposed to the temptation of a peace intended to be final." The practice of cosmopolitics is the opposite of finding a place of transcendent peace. It is a practice that Haraway (2008: 83) calls making "artful combinations" out of "speculative invention, and ontological risk" or what she later calls figures of "speculative fabulation" (Haraway 2016: 12). "Figures are not representations or didactic illustrations, but rather material-semiotic nodes or knots in which diverse bodies and meanings co-shape one another" (Haraway 2008: 4).

Leite (2014: 266) puts the question about imaginaries nicely when she asks, "how do we capture an inchoate, fluid, dynamic phenomenon that is simultaneously demonstrably collective and yet necessarily ontologically particular?" Figures as knots, knotty nodes, like the speculative way in which this chapter is written, are in line with my notion of imaginaries as non-representational assemblages, not (yet) representational, as Salazar (2012) usefully tracks them, but as a contingent and generative force on the way to a representation. My idea of an emergent imaginary starts there. Caribbean tropicalizations are creatively expressed materially in the flesh and used both materially and semiotically not to the demands of a dialectical expression, as Di Giovine, Swain and several others have developed it in tourism studies. Their work deals with imaginaries in their representational modes, as fluid poetic and political contestations and as shifting fields of power and meaning. I am trying to gesture toward something that runs parallel to these processes as non-representational, affective forces of intensity and potentiality, a sort of fluidity or flow not yet captured or contained: incommensurable incitements.

Thus, my chance re-encounter with a beer coaster, the mystery ship, and the intensities of the tourist ordinary in Wallaceville recall something of the generativity of a tourist imaginary as it instantiates itself as another "bad example" of the good life in Paradise, on a beach, in Belize. This is the way in which I read Taussig's line (in the opening quotation of this chapter). He exhorts us to "get in touch with the fetish." Mine is an effort to track a moving object in order to evoke something of the entanglements of its emergence and the emergency that animates these bad examples as singularities. This means tracking things and events as conjunctures of force, as assemblages made lively through tourist imaginaries (tropicalizations, images of Paradise) that incite seductive feelings of dream world relaxation and the affects of the good life to see what effects they have, what more they can do, and where they might lead if left unattended. The writing in this chapter is meant to nomadically and mimetically track Caribbean tropicalizations of a dream world Paradise resonant in the productive micropolitics of emergent structures of feeling that are generated through contact with my souvenir as it conjured the disparate scenes of a male model strip dance competition at a beach bar, a mystery ship that appeared in Wallaceville Bay, and the nervous intensities of ordinary, expat, and tourist life.

Seduction and shock are the affective forces that grow against the dream worlds of tourist imaginaries of a Caribbean Paradise and its nightmares. My Belikin Beer coaster conjured arresting images like the sudden appearance of Chas's Aunt Jemima in the middle of a public male strip show of Creole and Mestizo male bodies. Bodies like Jimmy's interrupted the different stories of everyday life with an overpowering and contradictory sensibility that literally "makes matter" of a nervous encounter with, and an association that erupts out of, the order of tourist imaginaries Belizeans and tourists use to lay claim to Paradise whether it be spectacular or ordinary. When affect makes its jump between the visible thing, the idea, and the socially sensible, it leaves a vibrant and kinetic-material trace of uncanny connections that begins to mark otherwise disparate states of arrest. This is the nervous shock of an unassimilated trauma and the giddy euphoria that drifts like a catastrophic dream onto center stage in light of the magical and profound seductions of my re-encounter with a beer coaster, in an uncanny place of encounters and becomings like Belize.

Note

This chapter was originally published as "Belize Ephemera, Affect, and Emergent Imaginaries," In *Tourism Imaginaries through an Anthropological Lens*, Noel B. Salazar and Nelson H.H. Graburn, eds. (London: Berghahn Press, 2014), 220–41.

Chapter 6
BELIZE BLUES

Find your blue.
 —Smithsonian National Museum of Natural History, Ocean Portal

As blue as devotion . . . blue is concentric motion.
 —Ellen Meloy, *The Anthropology of Turquoise*

Writing tries not only to accept the risk of sprouting deviant, but also to invite it. Take joy in your digressions because that is where the unexpected arises.
 —Brian Massumi, *Parables for the Virtual*

Atmospheric Blues

Belize, how blue it feels. It is not just one blue but blue hues and vibrant tones that press in on life, shivers of forces alluring enough to excite the skin and enliven the senses. Belize blue challenges language to encompass it. Yet there are just too many more blue sensations than there are words to reference them. Still these blues pull a social into focus in Belize yet register in ways that animate life beyond itself in spatial and temporal dissonant conjunctions, refrains, and scenes. Belize blues constitute an affective atmosphere (see Anderson 2009; Stewart 2010a), an actual materialization of high and low pressures, the dispersed in-fill of distances that is the light that gets lost in that space between here and there.

Atmosphere: *atmos*, something in the air, vapor, molecules, building, intensifying and releasing, brightening and dampening, breathing and exhaling into *spheira*. "Spheres," or bodies, enter into relational, gravitational flows of feeling that mingle in ways that are never finished or fixed in position, nothing captured, but are the push and pull, the flows and frictions of tactile acclimations.

In another register Rebecca Solnit adds to the way blue materializes as an atmosphere, as lost light and distance:

> The world is blue at its edges and in its depths. This blue is the light that got lost. Light at the blue end of the spectrum does not travel the whole distance from the sun to us. It disperses among the molecules of the air, it scatters in the water. Water is colorless, shallow water appears to be the color of whatever lies underneath it, but deep water is full of this scattered light, the purer the water the deeper the blue. The sky is blue for the same reason, but the blue of the horizon, the blue of the land that seems to be dissolving into the sky, is a deeper, dreamier, melancholy blue, the blue at the furthest reaches of the places where you see for miles, the blue of the distance. This light that does not touch us, does not travel the whole distance, the light that gets lost, gives us . . . the color blue. (2005: 29)

As lost light, blue scatters. It excites and activates things in the space between here and there. In a distance that you can never make up, blue is always "the color of where you are not. And the color of where you can never go" (Solnit 2005: 29); it instantiates things in an actual-virtual circuit that is never closed and is always indefinite. Blue activates and organizes incommensurate things, touching them, scattering them through a molecularization that changes them, and they are virtually infinite in their fragmented hues.

Blue is an activating quality of atmosphere itself, already there, the dispersed and mixed yet potentializing force of the haptic that presses in on the organizing of sensations of distance, duration, interval, vitality, rhythm, and desire, as we attend, attune, and attach to things as unfolding compositions "in social and cultural poesis" (Stewart 2008: 71). Emergent and enlivening atmospherics are the generative milieu of blue forces. These are unfolding moments of potentiality and are qualities felt as unpredictable, rogue or opportunistic somethings in an act of connecting up; that is what the surge toward an actualization feels like. What gets connected, or knotted into figures (Haraway 2008: 4, 88, 106) in moments of intersectional material-semiotic composition, provisionally actualize as objects, publics, and impacts.

Tracking such acts of composition means sidestepping an insistence on evaluative binaries like subject and object, the material and the representational, true and false, in order to take up a practice of sensing and evoking the colorful capacities of a milieu—a field of incommensurate forces that somehow hangs together for a moment as a consistency, a complicated intimacy of things that matter because they are thrown together as things happening: little moments, big scenes, nervous encounters, crazy intensities, odd vitalities, or uncanny hunches. So it is not the representational politics of blue taken up in different cultural or economic contexts that I want to dwell on here. Instead, it is a matter of attending to the charged-up tendencies and vitalities of abjection, seduction, happiness, and confusion that are fashioned out of moments of blue contact that compose themselves as a dense entanglement of affect, attention, and matter—slowing down "the indeterminate chaos of sensations enough to extract something from them that is not so much meaningful as intensifying [as] affective dynamized forces" (Grosz 2008: 3), serendipitous intensities rather than a system of signifying images anchored to a context.

For Deleuze (1997: 61) a milieu is more than just a contact zone (c.f. Haraway 2008: 4, 83–88). It is the different edges of an interface. A milieu is an autonomous zone of virtual-actual circuits, at once material and social—affectively infused intensities and trajectories that in their folding and unfolding enact new infrastructures of difference. They enact new modes of relatedness that affirm, augment, and keep open spaces of charged-up potentialities that index contact and, here, further the multiple occasions of blue's collective "presence" (see Massumi 2002a: 208–56).

By this I also mean that my writing is lured by a force field of seductive blue hues and tones that assemble into a milieu of transiting social, material, psychic strata—a transitional immediacy of real relations of desire, opportunity, and challenge. This milieu composes and consists, assembles and reassembles, in ways that are incommensurate with the telos of imperial prerogatives of globalized tourist capitalism as those are enacted in the wild zone of power on the frayed and fractured edge of empire in Belize, as it churns out the promise of plans for a "good life," Caribbean style.

For me it's a matter of attending and attuning to such a milieu of blue sensations. It is a matter of taking "a step sideways into what normally gets stepped over" (Stewart 2010a: 448) in the social analysis of tourism. It's a matter of being right where you are, only more intensely, in the eventfulness of various performative expressions of blue contacts—in Wallaceville, Belize, in the trade wind heat of the Caribbean, in a bustling tourist destination that is dappled in blues—and lingering in what matters in singular moments of blue contact. It's a matter of attuning to the alerted sense that something is happening—and then sensing it out (whatever *it* is) as an act of composing, worlding refrains, as an oscillation of strange forces resonating between something dreamy and disastrous (see Stewart 2010b).

In Belize, blue atmospheres encourage exciting and stunning Caribbean hideaway holidays that keep you buoyant while you ride a beachfront high. They are scenes of dazed and dazzled tourists wrapped in their loud, happy voices, smelling of sunscreen and sunburns deep in the seductions of a good-life bubble. This is the blue of tropical Paradise that seduces tourists with a desire for contact, with what seductively attracts as a "tropicalization" (see Thompson 2006). This is the happy, performative, industry blue of Caribbean tourism. You know this blue, you've felt this blue, if only in scanning the travel section of your local newspaper. It takes your breath away and makes your skin dimple with the pleasures of holiday sun, sea, and sand adventures. It is the global Paradise blue of the Great Blue Hole of Lighthouse reef—that ubiquitous sign of "supernatural" Belize—both awe-inspiring and mysterious in its instant changes of color: from the turquoise blue of the reef that defines the Great Blue Hole to the deeper than deep blue of an unfathomable sea floor sink hole, a "stunning jewel set in a coral ring" (*Destination Belize* 2003: 6).

Jacques Cousteau, that popular Ur-tourist-adventurer, was the first to explore the Great Blue Hole for us on TV, to reveal its mysteries and secret dangers, to map the "tranquil abyss" from his famous boat, Calypso, and so connect us up to the spasmodic forces of international ecotourism (http://www.belizeaudubon.org/?page_id=3603). And since then, avid visitors of all sorts tour "The Hole" and its mysterious underwater channels and passages. But stories are legion about its dangerous depths and the alarming number of careless unfortunate bodies lost, only to be found years later suspended breathless in some underwater dead-end passageway or tunnel. And all of this rubs up against Captain Morgan's Resort and Casino, whose very beach, they say, was the inspiration and original film set for Disney's *Pirates of the Caribbean*. And together the two create conditions tailor-made for true adventure, getaway holiday memories "that will last a lifetime" (*Destination Belize* 2003: 2).

Blue atmospheres also seize things up to become a local Creole tour guide's confusion and a local bartender's moment of endurance as they get caught up in circuits of active and reactive forces of official tourist training scripts that perform a perplexing difference between service and servitude. There is longing here in the way local life in this "little fishing village" has become unanchored and unstuck from its own common practices, modes of sociality. And so it floats in the super-charged-up tourist industry pressures—another tangle of associations, accrued strata of impacts, impulses, and reactions that enacts a deep ambivalence, ricocheting wildly between a buildup of deep sadness of loss and disappointment and the exciting possibilities of new dreams and schemes of "recovery," fed by the new world order magnetism of global tourism.

But this is the same alluring blue that indexes an expat couple's sad disappointment when their senior adventure in tropical living, once the temptation of a happy Paradise retirement plan, turned into a long-term attachment to a bad situation and they started crying some post-happiness blues (see Ahmed 2010). No one said anything about the downside of a life in Paradise or about the failed promise of a dreamworld, retirement good life. No one said anything about that unfathomable distance of a Caribbean beachfront tropicalization across which longing travels to finally lodge in the senses like a hideaway nightmare, as sad and slumped as the fading rush of a burdened blue light being pulled into nighttime darkness.

"Longing, because desire is full of endless distances . . ." Solnit (2005: 30) writes, quoting the poet Robert Hass. If blue is the color of longing it is because longing for the distances you can never make up is that activating quality of blue atmospherics. Blue, like longing, gets caught, dispersed in air and scattered in sky and water, and so never travels the full distance between here and there. Longing is always the space of desire for an object that is forever present in its absence and so it agitates in that endless open distance.

And in that space between here and there, light shifts, things happen, and things move, something is lost and dragged down into its molecular connections or is found but remains ephemeral and haunted. Nevertheless, something emerges momentarily as *something*, some everyday contingent sensibility. But something else escapes connection and goes rogue, becoming vitality in the reach that animates while drifting as an emergent line of resonant force in everyday sensibilities with the power to instantiate other things. Lost light may be ephemeral, but it always captivates us as it suggestively lingers on the wild edges of things. That is the blue of distance. "It is the distance between us and the object of desire that fills the space in between with the blue of longing" (Solnit 2005: 30).

Each scene of blue contact mobilizes into "worlding refrains," something Stewart calls "a scoring over a world's repetitions. A scratching on the surface of rhythms, sensory habits, gathering materialities, intervals, and durations" (2010b: 339). Refrains are an everchanging movement of interludes and fade-ins, gaps, extensions and fade-outs, reveries, vitalities, tangles of associations, accrued layers of impacts and reactions: all attunements that conjure stories generated through the inchoate impulse, feeling, tones, and affecting hues that bear an influence in Belize as waves of atmospheric pressures. And—like the endless repetitions of waves washing up on the Wallaceville beach that keep the form of the beach in constant flux—refrains are repetitions, agitations that gather as habits, accretions of intensities, sudden or sagging nascent forms, forces of chaos that become a rhythm without a regularity, a generative unfolding, "in-forming." Refrains create the feeling of what's around, the surrounding influences, the forces of various intensities, "affecting and being affected"—lived strains of action and reaction condensing as capacities not yet bearing their own form, but "in-formation." In short, these Belize blue refrains are atmospheric compositions: worldings.

Such blue refrains, scenes, and intensities, generated as circulating atmospheric forces, become the backdrop of living in and through a shifting beachfront pressure system of highs and lows. These circulating forces are perpetually forming and deforming Belize bodies as qualities not in their own right but as intensifying mal-adaptations that swirl, build and subside, that are strong enough to pull bodies into awkward, fussy alignments with other bodies, worlds, histories, horizons, rhythms, and ways of being in the climate of an eventful, crazy, alluring, and enduring postcolonial, late liberal, beachside saturation of the senses for better and for worse. Belize blues distinctively reveal stunning Caribbean tropicalizations in which official international tourist seductions, national, local and expat cultural expressions necessarily participate as bright vitalisms that ride the cusp of the incommensurate, enabling lives to become otherwise.

But here, blue matters beyond its power as cliché, beyond the capture of life potentials, and so it is always way more than any representation of blue could ever offer. Blue matters in its movements, tones, and blends, in the unregulated rhythms of its shifting vibrant hues. Tropical blue is a relational force that produces sensations, affects, and intensities that forever exceed whatever blue's actualized determinations attempt to fix: blue becoming "all too blue" (e.g., Massumi 2002a: 208–56). And it all becomes much more than the wager on which a dreamworld Paradise could ever promise to pay out or what a local future, as an inflection of a national economic assurance, or the bliss of a retirement rapture, could ever deliver. Imagine blue as a transitional force forming as a milieu that catches things up in the contact zone between here and there, that urges attention to its resonant disjunctive dynamism, to its creative potentiality, its uncanny, interstitial and emergent liveliness, its excess. This is also how blue emerges as an event in that nonlinear space between here and there—sensations, a series of connections, an unfurling moment when something begins to actualize and then something begins to instantiate itself as part of a scene, as a color of a social composition.

Tourist blues, local blues, blue expats: three instantiations of life that, when we pay attention to light, compel those lives to some empathy with blue. My writing attends to these interlinked registers of blue contacts and emergence, selected out of countless, potentially possible moments, situations and scenes when a sense of something happening begins to surge up. How do blue atmospheric pressures of light spawn multiple micro-worlds, animate attachments, entanglements, and unravellings, and act as the intricate situation-scripts for how forces become lodged in bodies as vectors of social possibility, the back-form of potential's unfolding and circulation? How do forces of blue propel and compel, attract and detract, convoke and provoke things in the making on a sunny Belize shoreline?

Here I evoke something of the qualities of blue intersections while tracking its emergent, world-making force through everyday life in Wallaceville, Belize. So, it's the daily rhythms and attachments to scenes and objects of desire that sustain life there that I turn. It is to how blue takes material and imaginative form in the way it pulls an assortment of forces, events, sensibilities, and materialities into alignment to become nervously generative of the textures and rhythms of social life in a place that has gone crazy for tourist development—for better and for worse (see Little 2010, 2012, 2014). What are the emergent political imaginaries and affects of blue compositions that might become thinkable and sensed through a sustained engagement with the beguiling and recalcitrant materiality of that good-life feeling, tone, and hue attending to light's sensitive, trapped lines to feeling blue?

Paradise Blue

That blue happiness: picture yourself somewhere on the Belize coast, on the barrier reef, among the teaming corals of the cayes of the Caribbean Sea. Imagine the bluest of azure blue waters, a powdery white beach, a smooth fast boat trip to Lighthouse Caye and diving at the Great Blue Hole or an infinity pool, a hammock, and the rustling rhythms of tall leaning palm trees swaying in a gentle breeze. Feel the warm water touch your skin, caressing waves lapping your body. A vibrant blue sea matched only by the bluest of blue skies. It is an amazing blue that saturates you, connecting with a solar radiance and a sticky humidity that hangs on you; it dampens some things as it enlivens others. You feel immersed in something pressing in on you. You breathe it, in big drafts that taste a bit salty, gritty, a bit burnt or fragrant; it's all a bit mysterious. A collection of sensations alerts you to something mixing and to something emergent in the mix that charges you up, it changes you. It stirs your imagination and you smile: it's that blue, conjuring good-life attachments to Paradise, with fresh hibiscus blossoms in your hair, an exotic rum drink in your hand, and soft Bob Marley tunes on your blissed-out soundtrack. The point of your trip is to stay in that seductive, dreamworld, Paradise bubble and the affects it encloses. The natives are "super friendly." The village is safe and the good life rolls out without a hitch. Now this promise of happiness is on its way to being realized, actualized against Sara Ahmed's (2010) better intuitions, at least for the moment, and your smile broadens and deepens while you too begin to unfurl and unwind.

You may never have been to Wallaceville but if you have seen any of the extraordinary pictures used to sell Caribbean tourism you have a tempting and seductive image of the place. You already know it, like it, and want it even before you think about it as your sight lines meet the "infinity" pool's smooth projection into the sea and to the horizon. No interruption just endless extension. Beautiful bright sandy beaches, the warm blue water, the brilliant blue sky and sunshine, you can feel the soft warm trade winds blowing across your body. It's relaxing and dreamy. You can swim, sail, scuba, and snorkel or spend your "down time" in a hammock under the shade of a palm tree sipping beach drinks and taking in the splendor of nature and local culture.

The Belize Tourism Board advertising brochure says that Wallaceville was once a "sleepy fishing village" that is now also "an exciting tourist destination, where change has taken place without losing the original Belizean culture. . . . A sidewalk meanders through the village along which you will find a variety of gift shops, beachside bars and restaurants specializing in everything from local Belizean dishes to more exotic Caribbean cuisine. . . . Once you have rested you can enjoy the casual nightlife the village has to offer" (*Destination Belize* 2003). There you have it: a warm, sun-drenched day and glorious evening in Paradise. It's smooth and mellow and intoxicating, it's what you have always wanted in getaway pleasure, in a pirate hideaway cove advertised across the globe as "mother nature's best kept secret."

But it's all more than that, too. Wallaceville is a place that tourists have been getting high on for a long time. The stories of hippies discovering the place in the 1960s are legion, but its discovery also has a longer history that takes us back in time to the eighteenth century of pirates and privateers and later to the blue indigo extractions of the nineteenth century logwood trade. All of these stories of discovery act like sediments of romantic discourse that stimulate cultural images that act as fossils do, preserving a radioactive quality of original contact to be tapped into in the contemporary tourist moment of wish image encounter (see Benjamin 1999).

Wallaceville is awash in blue tropicalizations. It's a place where a body can't help itself as it mixes bliss with just a bit of dread about how good it all feels and just how long such bliss can last (see Little 2010). It has a reputation for good stories, food, ganja, cocaine, Rasta sex, friendly natives, and booze all tied to a beach party ethos, but in a relaxed, sunny, and laid-back mode. It's a place to detach, which is the tourist's prime directive. It's a beachfront in vitalizing blue where the body surges, drifts and dreams, gets sidetracked, indulges, falls down, crawls, gets up slurring, indulges some more, laughs it off then hits the wall, regroups to do it all again, or beats a retreat and gets out. This is a place where the tourist body knows itself in states of intensity and vitality, exhaustion and renewal.

But the buoyant wonder of this blue is almost always too good, and the good life in Paradise it conjures is a delicate balancing act of advertised Paradise vacation elation that rubs up against a local duty to service Paradise and to some over-the-top, urgent need to meet the demands of the long- and short-term tourists by reproducing advertised tropical happy objects. And so the tourists come, the cruise ships come, and so does the big-time crime, the drugs, the land-steals, the corruption, the violence, and too many mysterious changes to village life to count—like the new demands on the local, the new good-life rules about pets, garbage, language, dress, and comportment, made to accommodate tourists and leave them with a lasting impression of how wonderful Paradise is. The collective sensibility of living up to the advertised blue hues of the Caribbean tropicalization, now in hot circulation globally, drives every local more than a bit crazy. Wallaceville locals, those born and raised there, see it all happen before their very eyes, this blue liveliness coming into play in new scenes, habits, and in moments when things throw themselves together into something that feels like a situation and the object of ordinary attention.

These instances of blue are placeholders that magnetize heterogeneous fantasy investments in Paradise adventure. These blue sky and azure blue sea tourist industry worlds are sites of collective excitations that organize as moments of tourist engagement, blockage, and desire—states of emergence as lines of force immanent in that activated infinite reach of light dissolving into that atmosphere constituted in that space between here and there. What emerges in this milieu is not just a good-life promise or a commodity mystification but an excess of all those moments of tourist vitality, potentializing forces, singularities not representations, weak, unstable significance. What emerges is not so much signs that carry meaning, but co-constituting textures, impulses, and densities that pick things up as they move episodically through bodies in touching moments of aching topical beauty that lodge in moments of crazy beachfront fascination settling into some life of the senses.

Local Life in Indigo

For Creole locals the blues of Caribbean tropicalizations feel like the ambivalent, fraught, and dissipating feelings, tones, and hues of tourist service and a deep and abiding sense of historical servitude mixed with a new contemporary tourist industry hustle and laughter. These are blues that fill Belize coastal villages with the dynamic sensations of possible fortune and futurity, and that back-talk local hardship and despair, local circumstances and traditions, into something that begins to recalibrate the social, the local, and the sovereign. That's one way that local blues blend in with that ubiquitous blue of a globalized Caribbean imaginary (see Little 2014).

Take Ron, one of Wallaceville's blue ribbon tour guides. And he shows me the ribbons and medals and plaques to prove it. He wins National Geographic and Belize Audubon Society awards for his knowledge of local nature and culture, for his hospitality and for the professionalism of his adventure wildlife tours. He has worked with the Smithsonian Institution too, helping adventure ecotourists to "find their blue." He knows his way around Belize and can guarantee an experience of a lifetime for tourists and adventure naturalists. His reputation is worldwide as the reliable Creole exegete. And he knows Belize nature. He and his wife made enough to build their own modest tourist cabana operation several years ago; their business is barely making it now that the big resort tourists are replacing the day-tripper backpackers, if anyone visits their place at all these days.

Ron is at his wit's end with all of that, but it is still *a living* he understands. And it is hard work dealing with *the industry* and the tourists with short attention spans, mostly looking for their next meal, beer, and bathroom break. When he is not leading a tour, he trains tour guides through the National Tour Guide Training Program of Belize. He has worked doubly hard to help the younger villagers become effective guides, but he is close to scrapping the whole project, a dedicated professional lifetime of work swirling down the drain. These days Ron mostly gets *rass* from the trainees. He says that they won't buy his lecture on *service not servitude*. *Work hard for the tourists means working hard for the nation and that means working hard for you*, Ron says. *Guiding is a good life. Tourism makes things happen in Wallaceville*, he says. And his training sessions are performances that open onto the shifting grounds of a crazy future for postcolonial circumstances in Belize that he has no choice now but to tap into and ramp up. Ron's greatest success is his example. He says he is a *winner not a loser* and if he can win the kids can too. But his trainees laugh bitterly, and their laughter unsettles the official happiness tour guide scripts that Ron passionately protects as the keys to success. *They don't get it and they are not buying it*, Ron complains.

For Ron, there are two ways of squaring this new tourist world disorder. It's about knowing the difference between service and servitude, when for those he is desperately trying to train it's all servitude. *Excuses*, Ron says. *They're lazy*, he complains. *Too much TV, no ambition, no sense of commitment. They get it from their parents. No one works anymore, not really. They want everything given to them. They're all a bunch of lampas* (the Creole word for lamp post suggesting the boys just hang around like lamp posts, or on them). Ron knows that Stretch *hangs* at the beach with the tourists and hopes for easy money selling ganga, taking in the tourist parties that he may be invited to. Maybe there'll be a free trip to Europe if he attracts a *sugar momma* or a fancy job on a yacht or at one of the big resorts. Wait on family money from NYC, Chicago, Miami or LA, *gangsta* clothing, and bling. *Gotta look good*, Stretch laughs. *Man, Ron just kills the vibe*, he grumbles. And Stretch may have a point, his friends say. Everyone respects Ron, his dedication and ambition, but the real models for getting ahead are the politicians who use or invent the rules in order to make secret property deals and drug deals with who knows whom, from who knows where, and getting rich doing it. And when they roll into town their success rolls in with them in their new SUVs, with their new big boats, new fashions, new friends. *Get in, get out, have some fun*, Stretch insists with a smile.

Ron's life of service is Stretch's servitude. *Skip it*, Stretch says. *Like da song go, man, "Don't worry. Be happy,"* he laughs. Ron's true-blue work ethic no longer attracts those who see another world of Paradise happiness emerging while performing versions of themselves as tropicalized happy objects, *fo wi di real Creole cultcha*, Stretch insists in Kriol. And the *once in a blue moon* success stories, the lucky ones who have hooked up and into this erratic and discontinuous new world *whatever*, have become the new impulse examples for local kids like Stretch. *Find a beach wife. Be like a tourist, that's where the easy money is.*

Or take Sweets, one of Stretch's best friends and someone he often tries to emulate. Locals have been shocked and seduced into this new world of chronic capitalist blur, fueled by Belizean tourism, and touched by the threat and promise of the tourist spectacle of a good-life Paradise. Sweets knows to stay in *happy-go-lucky* tourist frame. *Life is "Sweets" that way*, he chuckles. And he knows that locals out of costume, out of image, and out of frame are a threat: another arresting image (Stewart 2003a) of local impact. It's as if locals are literally touched by the scene and make themselves its convulsive possession.

But this all makes Ron gloomy blue as he worries deeply about the transformation of his cherished little village that has gone crazy for an unfolding, episodic, moment of unruly impacts that inarticulately blunders about on the margins of global control society—no clear plans in sight, only the individual willingness to improve by going with the flow, despite the disappointments, the insults, the rip-offs, or the humiliations. And now it's the cruise ships. I caught up to Ron one evening after he had stomped out of another community meeting with the representatives of a leading European cruise ship line. They now have big plans for a beautiful little caye located just down the coast from Wallaceville. They were in the process of turning the caye into a private investment, international cruise ship terminal and Disney-like tourist village: marina, resort hotel, condos, soft white beaches, lounge chairs, and big colorful sun umbrellas, you name it. Ron was more than angry. He was almost in tears and shaking as we walked, feeling deeply *disrespected* because of *the man's rass*. The cruise ship rep had just laid into him during the community meeting of nearly three hundred people, where Ron and other locals who were adamant in their opposition to this development project were given the chance to lay out their concerns:

You just don't get it do you! the Man spit, smiling hard at Ron.

The Man said things like: *You have to separate your emotions from the facts, Ron. Read the EIA. We have the facts so calm down*, he yelled. *We addressed all your questions. Do your homework Ron*, he commanded.

Once you look at it rationally and understand the improvements in tourist industry investment we will be making, you will see the long-term benefits to the community. That's when I'll say, "Welcome aboard Ron," he said in a bruising blue tone of dripping condescension. *Just because you are offended doesn't mean you are right Ron. Next question please.*

But Ron is a nationally trusted naturalist with impressive international credentials, and he is usually a respected voice of local authority and reason especially when it comes to the world of tourism. He knows what's up and so do the ecotourist NGOs and some of the other local business people he talks to. He has seen it all before, with the cruise ship Tourism Village fiasco in Belize City. His urgency was matched ruthlessly with industry disrespect and Ron felt all of it:

> *I don't know anymore*, Ron says grudgingly. *Maybe Stretch and Sweets got it right, Man. Maybe it's all servitude*, he admits sadly. *What am I teaching them for? Who we working for? My guess is that these guys will make some kind of public relations gesture to appease the community, maybe some contribution to help with the waste disposal issue in the village, build some schools or hire more local people, pay off the rest. The deal's done, land's been bought, politician's been paid off, corporate wheels are turning. You can't undo all that, not now, too much involved. This meeting was a useless thing. It's their game now*, he laughed harshly.

We walked up the village sidewalk in silence with the sound of the sea on a warm soft breeze, the sun setting, the darker blue of the encroaching nighttime touched Ron's blue mood and he turned bluer than blue.

Wallaceville comes together for Ron through a relay of discontinuous and harsh moments like this. In a flash he can trace a litany of such moments of confusing ambivalence when his trust in a service ethos and practice rubs up against an ever-expanding atmosphere of out-of-control servitude. That's when things either successfully come together for him or things just fall apart, accrued layers of substance building, ebbing or flowing like the waves on the shoreline as Ron struggles to find his blue. Ron is ready to admit that the choices are stark: service or servitude. But while the two nervous edges touch to form a circuit and a determinate context for the big story, which actualizes as the good and the bad sides of local political power and tourism capital forming a low pressure zone of intensity that forms as a milieu, they are only two edges of an enfolding interface.

Ron's stories of service and servitude grow palpable—conjuring rogue trajectories of potentiality, the unqualified possibilities of excess that are those infrastructures of difference forming the contingent space between here and there. This might be the something blue in what fails to work, or gets lost in things, or that might be waiting for a chance to occur—something nascent in the atmosphere that acts as a pressure, as image touching matter and color touching us, that keeps open spaces of charged-up potentialities and furthers the multiple occasions of blue's actualizing as a presence. But it's that feeling of a new servitude that rubs Ron the wrong way and puts him in contact with the stories about his own family history when eighteenth- and nineteenth-century Creole woodcutters were forced into collecting logwood for the British dye industry to make indigo blue: a color that now indexes the sad and sordid way things feel today for him.

Still there is some odd voluptuous pleasure in Ron's blues, because it is in moments just like these that Ron *gets lost*. That's when he disappears out to his family caye. He never feels better than when he beats a retreat and *escapes*. He sails his boat, his social life either before or after him, races across the shifting blues of the sparkling sea under a brilliant blue sky suspended in the beautiful solitude of a shifting distance. It is here too that Ron finds his blue: in the introspection that provokes his memories of growing up on this water, fishing with his grandfather, the weeks out on the reef working, fixing up the family caye that is now his. The feelings that this transitioning from land to caye conjures is something like an *aching joy*, Ron says, when the blue is deepest on the horizon and the clouds are *doing their wispy fleeting thing*; it's one of those somethings *easier to sense than to explain*. Just before night falls, Ron's blue is that last quantum of visible light to pass through the atmosphere before it scatters dark. It draws Ron into the skin of the world feeling the tidal pull of light on water. That's when Ron can breathe again, the breath of someone brooding, to be sure, but also of something inexplicable, beautiful. It's a place where desire floats free on changing currents that shape his life, every bit of breath and blood.

> *They say that the Maya used to mix clay and indigo together*, Ron says, *to find a blue that the dying could take to the afterlife. There is something amazing in that connection, don't you think?* he asks quietly, his eyes and mouth half-forming an expression of amazement, like a thought in the making, not there yet but still enchanting.

Alberta Blues

When the name for a color is absent from a language, it is usually blue.
—Ellen Meloy, *The Anthropology of Turquoise*

> *The blue of it*
> *moved by blue things.*
> *And then the blue people*
> *whose vision has been injured*
> *or altered.*
>
> *The cause of injury is emotional*
> *the extreme weakness of organs*

> *and the inability to recover.*
> *A deep blue bruise.*
>
> *The same blue that delivers hope*
> *can also deliver despair.*
> *To find yourself*
> *Trapped in blue*
> *no matter what the hue,*
> *can be deadly.*[1]

But Ron isn't the only person with a case of the ambivalent blues, although for Doug and Gaile things didn't start out feeling that way at all. It was all vibrant happy holiday blue that seduced them to Belize. But it can become such a heartbreaking faithless intensity. Their tropical early retirement "good life" now sags with disappointment and acts as an incitement for their unhappiness and regret as the hues of this seductive daydream blue began to blend into the monotonous hues and tones of their daily life in a disconcerting way—conjuring a dream gone bad and attachments that now feel frayed, a dreamworld optimism turned dirty blue, mean and nasty—yet they still have some irresistible cruel attachments to the things that will probably kill them. This seducing scene rubs their senses raw and beckons toward a wretched fade into some contingent combination of sensations that make them both feel blue. It indexes an expat "situation tragedy," a post-happy moment of sad disappointments (see Ahmed 2010; Berlant 2011).

Doug and Gaile really felt like they were into that something wonderful which Belize blue promised. They came from Alberta, out of the oil patch, left it all behind, packed their hopes and dreams for Paradise and blew in on a wisp of a warm Caribbean breeze, light and fresh, like a high-pressure system. They were looking to find their blue and thought that they had. But it has all become a *bad trip*. Dripping exhaustion in the heavy humidity, mostly they feel a drooping frustration mixed with sadness, shock, and incredulity. Their talk these days is always about some local tragedy or about someone recalling a tragedy that recently unfolded and it all conjures local hopes abandoned and dreams once glimpsed now lost.

Like the day that their local Creole neighbors just to the south of them warned Doug and Gaile not to let their building developer destroy their handsome stand of shoreline mangroves. But they needed a dock for their big new boat to go with their big new *beach palace*. Besides, who knew better about local seaside ecology, the locals or the experts? Everything would be just fine, their developer said. But now that their third boat and motor are gone, stolen yet again, they don't need the dock. They don't even like to look at the dock. Iguanas, attracted by an old unpruned cashew tree, peacefully occupy it as if moving to the rhythms of a less fractious and sunnier time. Their mess fouls and agitates the once handsome structure that now, ignored, rots away stinking of green. *A sore spot*, says Doug sadly. *An infected spot*, Gaile laughs bitterly. The mangroves were cut down to put in their dock and the docks of several of their neighbors to the north of them so now their once clear lagoon water has turned into a murky, reeking mess of sulfur smell and heaps of floating garbage that congeals in rotting piles around the dock and shoreline.

This bolt from the blue is almost more than they both can endure when they look out at the smelly overgrown chaos that feels like it parallels their lives in Paradise. Their lives have become a tropical low-pressure zone and the best Doug can offer is a bitter note in a growing expat refrain, *You know*, he says so sadly, *we're into somethin' now we really can't get out of*. Trapped. Growing paranoia and fear. Substance fatigue and worn out nerves. A troubled, other side of Paradise they didn't see coming. They desperately want to sell and get out, but since the '08 market crash no one's buying and besides, where in the world would they go? They can't afford to leave, even for a break. Leaving the house empty, even with a guard around, means a break-in and who could they trust with all of their stuff anyway, a tourist? It's all a profound disorientation. A slumping nightmare, Paradise is all about enduring now as some emergent structure of feeling that they can't shake at all. This is the poesis of an ongoing, unendurable present built on a longing for that dreamworld bubble that has now popped—and so things become part of that unfolding moment, a state of emergence that makes things aggressive, unsteady and fugitive, the potentializing state of emergency that goes for the everyday for Doug and Gaile and so many others just like them.

So now Doug and Gaile suffer daily through their tropical depression. Who knew their dreamworld retirement investment would fall apart so easily and so quickly? They have tried their level best to fit into some village routine that worked on them originally as the promise of a generic Caribbean cultural seduction, but it turned out to be anything but. They see it all now as part of their sad refrain, a scene that shocks the growing and nervous exhilaration of local consumer tourist culture in Belize. Along the main road for as far as you can see, new expat neighbors are building happy-life gated beach mansions or timeshare condo communities, outfitted with golf carts, in internationally financed mega-projects so that there is no public space or comportment left. Cozy holiday and retirement cocooning are where forms of Paradise living have become tactile, and the bubble of fantasy life born of commoditized "local Caribbean culture" grows sensuously vibrant in the circulating impacts of a tropical dreamworld haven and settle in as the dreamier side of this arresting image of Caribbean retirement life.

But, as Doug and Gaile know now, it all places heavy demands on an image. The holiday-retirement house in Paradise was meant to guard against the *outside world* with its wild scenes of global crime, chaos, violence, terror, disease, and decay. Doug and Gaile thought that the time was right to escape to this hideaway haven, to revitalize and find renewal and a new purpose in life. Life's little pleasures hooked up to the big picture of a local Paradise that settled into a dreamy connection with things, a resurgent image-effect of tourist pleasure mixed with a retirement master plan.

Vitality and happiness unfold as if naturally and effortlessly—but, as they finally felt it, anxiety, fear, and the shoreline ebb and flow of bitter disappointments are the unsteady grounds over which happiness flows, breaks, and founders. That's when the big-picture dreamworld implodes under the weight of its own embodiment and plays itself out to the point of nervous exhaustion. A dreamworld beyond the pale becomes a nightmare (Stewart 2003a). That's when the dream is confronted with its own contradictory and excessive strivings for pleasure and that's when Doug and Gaile began to ask themselves why they were trying so hard to relax. That's when the panic began to settle in over the fear, and disappointment felt like a growth industry, like the way Massumi describes good food as being the foretaste of heart disease (1993: 9).

Their beach side hideaway in Paradise first led to a regime of healthier lifestyle choices and a chorus of new therapeutic routines that were supposed to set the controls for beachfront happiness. They decided that it was best to keep busy, so they put more effort into stabilizing the dream. They tried to get involved in the community, to help build a cleaner village and proposed plans to deal with vermin, stray dogs, and garbage. They hooked up with a local NGO ecotourist, save-the-reefs group too, that Gaile says they poured a fortune into. But it folded up and vanished some time ago along with their investment even though they still get a year-end financial statement and a "thanks for your support" from their main offices in Copenhagen. And, besides, most of the locals never really got involved in any of these community plans anyway, other than the kids from the village school who showed up on class projects days sponsored by the village council or the stalwarts from the Chamber of Commerce or the Belize Tourism Industry Association (BTIA) to help out one way or another.

Doug and Gaile built their beach dream home on the intensity of a seduction that had become seriously frayed. After their second boat and motor vanished, they installed some sophisticated and expensive surveillance technology, this time guaranteed to keep the local thieves out. But it didn't. They used to share information about break-ins, protection, cell numbers, who to trust and not to with other expats at their yoga classes or at the Soft Touch Thai Massage Spa or on the recently created neighborhood Facebook page. And they used to play horseshoes with the locals on Saturday afternoons when the Sunset Bar was still around, making connections with them and feeling good about the positive contact; but even that was a niggling reminder that they were surrounded by a nervous otherness that was never quite assimilated or domesticated, no matter how strong the dreamworld image of friendly locals was.

But their involvement meant that they were never able to escape the cruel world of local corruption and fear just outside their beautiful beach house and that felt like a betrayal. They know now that you can't buy land or build without some Belizean ripping you off or stealing you blind. No local honors a commitment; contracts are useless. Don't go to the police, things just get worse if you do. Finally, all sorts of fears—drugs, thieves, corruption, chaos, and unseen dangers, local looks—swelled, bent, and stretched things out. And there they are now, Doug and Gaile, right back where they started, nervously weighing their lives, as the future grows tense and tactile with anxiety and while their abused bodies index sad and sagging lives that have become a "cruel optimism . . . [that] exists when something you desire is actually an obstacle to your flourishing . . . when the object that draws your attachment actively impedes the aim that brought you to it initially" (Berlant 2011: 1). This is the moment when their Paradise slipped them into a blue funk.

That's when Doug and Gaile started to *lay low* and that's when they started hanging out more often at the local bars with the other expats just like them, forsaking the therapies and the community investment projects. Now when they think about their place in Paradise it's like riding the ups and downs, the heat and chills, of dread and pleasure: life reduced to atmospheric pressures. So stick to conversations about the weather. Don't get involved. Make the necessary master plan adjustments. *Don't panic. Take it one day at a time. Don't forget to breathe. Get your sleep. Try to relax.* But the specters of Paradise keep reappearing to haunt the dream, like some abject surge that points to a troubling vulnerability haunting the deep blues of a Paradise depression. It's excessive and virtual, nothing you can put your hands on or completely figure out, and it makes both Doug and Gaile sad, nervous, angry, and fearful, but not all at the same time. It's all mixed up with the daily highs of afternoon gossip get-togethers, shopping, and then on to the daily happy hour evenings ducking for cover with friends just like them. They are all oddly located in the ongoing nightmares of Paradise, feeling their own blues (see Little 2010).

But what now? It's in generative moments of ducking for cover when things get caught in active and reactive circuits, when the sad and sagging trajectories that once held so much promise begin to shift in feeling and tone that Doug and Gaile begin to sense the dissolve of a bright blue tone into another much darker one. *That's life*, Doug said one afternoon when he was recounting how it was that bad things always seemed to add to other bad things in a way that he could not make sense of anymore. For Doug and Gaile, Paradise life has become an endless process of getting themselves into something or trying desperately to get themselves *out* of what they got themselves into on the seduction of a dream and then it's another thing. These are the moments, Stewart (2008: 72) says, when "something that feels like something throws itself together"; when Wallaceville composes itself for Doug and Gaile as an uncertain and fretful unbecoming.

Still, the rumors about tourist development proliferate at a panic rate locally. Now the rumor of cruise ships coming into port and flooding the village with thousands of tourists at a time, with no infrastructure in place to accommodate them has become reality. "We are knee-deep in shit of one kind or other," the village council chair says to me one afternoon after a long troubling council meeting in which he and the other members rehearsed their own survival strategies while considering the imponderables and contingencies of local life heading out of control—yet another register of nervous agitation fed by conspiracy theories about those they know who are *connected* to the *cruise ship people* and those who stand to win or lose on future land deals, political influence, and the new tourism village that was built but that stands still mostly deserted, built on an agreement made by *who knows who from who knows where*. And all of a sudden, the image of Paradise becomes completely indistinguishable from the commercial globalized world of the tourism industry mixed in with sordid forms of local corruption.

Again the point is how the virtual and the real have collapsed in on each other, their difference no longer relevant in the state of emergency that is rapidly forming as a generalized condition of Paradise capitalism on the margins of empire, teasing and encouraging these new forms of life transformed mercilessly into strategies of survival. But what also needs attention—along with the big system explanatory critiques of commodity mystification, cultural politics of class and race, or the global inequality of the advance of tourism capitalism creating the state of emergency that Belize is becoming—is the insensible *what else*, the realm of possibility within the praxis of social norms and affective intensities that may release other modes of being out of matter forming in the interstices of blue, for better and for worse.

For better or for worse, life became a series of ill-tempered, badly timed moments of social collision and stale, unhealthy routines that have kept Doug and Gaile feeling off-balance, worn-down, and out-of-synch lately. These days Doug and Gaile feel more than a bit blue. It feels now like the village holds secrets that will never be revealed to expats like them. It's all about local life going on without them and beyond their control. And all they were trying to do was fit in and help out. But here they sit, half-numb guzzling two-for-one happy hour rum drinks at a local expat beach bar, nuzzling the facts but mauling the truth the drunker they get, Gaile softly slurring the lyrics to an Eagles' song as if the words figured her life out. She is squinting past me, at her bar-side familiars, blurry kin as kindred spirits, just like them, and with a sardonic smile she sings along with the others to their new theme song, *They call it Paradise, I don't know why. You call someplace Paradise, Kiss it goodbye.* Gaile looked at me and finished what she knew of the song as if it really meant something to her, like a forecast she wishes now that she and Doug had been able to read better than they did. *Some rich men came and raped the land. Nobody caught 'em. Put up a bunch of ugly boxes, and Jesus, people bought 'em. And they called it Paradise. The place to be.* And the title of the song, "The Last Resort," doesn't help matters much. It seemed to reach into the assembled expat nervous system of the bar like a crazy kind of sickening impulse that they all feel intensely if not all at the same time or in exactly the same way. They are all deliriously riffing on some vibe of last resorts. They each have their ups and downs and that means that there are some who can laugh at the lyrical ironies. But this time Doug and Gaile just look at each other, drinking and drowning from the inside out. Gossip, another rum drink, and the bar itself a miserable lifeline of connections, worn, leaning a little, and a lot worse for the wear and tear of this daily ritual, the moment stabilizes the grief, but only for the moment.

 They feel local eyes scanning them daily and sense the crazy changes that are transforming the image of Paradise they *bought into* for some fitful, cartel-driven, corrupt and out-of-control chaos. *They're all thieves and drug addicts*, Gaile rasps, talking about the politicians, as she lights up another Colonial, struggling to keep her sweat-soaked body perched on her bar chair while finishing up her early evening drinks, before she and Doug stumble home feeling flush-faced and soggy damp and tanked again, preparing themselves for yet another bumpy night to the echoes of street laughter that they are sure is mockery directed at them: it's all the background soundtrack of popping daydream bubbles and that encourages another hit of cheap rum, what locals call *blue ruin* that sours into something resembling horse piss, until smothered by the gloom they both, ragged and bruised, fall into a broken slumber. It's a sad joke, and they feel like they are wearing it and find little confidence in the fact that they are not alone.

 And that's another story of Belize blues, and those are the facts. But the facts are not the truth of this story of Doug and Gaile. I think, as I scanned their faces, these tired Canadian seniors, now lost in the twisted ebb and flow of a Paradise promise, lost in that space somewhere between here and there, that I can imagine the truth of this story. I imagine it lying in the thick, overgrown green mess of that dock, its broken promise screaming diminishing expectations. Doug and Gaile for whom winning at local life happiness in Paradise meant everything. A couple who had talent and a positive outlook on local life, with their forgiving happy nature, they had the potential and the skills to help the community, help make life better for everyone, keep things local, honest, and happy, and so, get along. But they didn't. And they didn't because the seams of that Paradise dream were worn and torn by disappointments and odd expectations. And they didn't because sometime in that rare moment after making love, smoking their Colonials in some cool blue smoky silence while listening to Mr. Peter's Boom and Chime on the radio next door, that music so true yet so distant, they glanced at each other and noticed the lines growing heavier around their eyes, felt what the others felt, bodies burdened in deep blue sadness. And like all those other expat couples their gaiety and light-heartedness were as unreachable now as their stolen boats and broken Paradise promises. And they didn't because they had the blues, felt them deeply. Their adventure into the hopes and dreams of Paradise blues left them with the feeling that being so seriously up for an adventure of a lifetime had cost them just too fucking much. All that is left now is some ragged remnant of themselves on a beach in Belize. When I last saw Doug and Gaile they were staring hard into the bottomless well of a rum bottle while sitting on the porch of their big worthless house. I caught their wry smiles against the blue shadows of the Maya Mountains, and I knew that this was another of those burdened moments of clarity, feeling the vast gap between the facts and the truth of a life in Paradise that will not easily be forgiven and never made up.

It is the atmospheric pressure of that space between here and there, created of that interstitial distance, that milieu of affective intensities; it's the blue skin and the blue lips of a nervousness jumping from feeling to fear that feeds Doug and Gaile with an aching sadness becoming their dominant mode of attention. The blues happen when you forget how to breathe. Here in the generative modality of a discontinuous assemblage of incommensurate yet potentializing sensations something like blue feelings and impulses throw themselves together, composing a kind of rhythm-tone to a desperate journey of connections that move for Doug and Gaile from blue dusk to a nightmare darkness of despair, from those happy blues and early days in Paradise to the cruelty of sad and sagging attachments—and that's when some *whatever* jumps into form to feel like something, for better or for worse.

The Light That Gets Lost at Its Distant Edges

> I used to wonder why the sea was blue at a distance and green up close and colorless in your hands. A lot of life is like that. A lot of life is learning to like blue. . . .
>
> —Miriam Pollard, *The Listening God*

Scenes of blue throw themselves together as potentializing forces creating moments when assemblages of incommensurate things, exchanges, linkages, sensations, and publics compose themselves as atmospheric pressures, as a co-constitution of things blue that begins to feel like something good or bad or odd or rogue, or eerie or completely off the scale. Each blue scene is an attunement, a "worlding refrain" (Stewart 2010b). Here Wallaceville is a space of unfolding where some active generativity of things takes hold and forms as a state of emergence in a state of emergency that hits the senses and makes bodies jumpy, unstable—makes them shift, just as they may begin to feel adaptable, resourceful, ideal, useful or not. Each scene feels like something bodies get themselves into, comfortably or not, for a short time or for what feels like an eternal hell. It's a full-on enthusiastic immersion or a nervous twitching that skitters along the edges of things, composing as scenes of impact. And then it's something else, like a dream or a kind of bailing on a dream: that feeling of being abandoned by the world or seduced by those who become the "trust investments" or examples worth emulating. Things all depend on the lively entanglements of affect and attention, matter and flesh: a virtual to actual move that always leaks.

This writing is my augmented worlding refrain, a materialist semiotics attuning itself as a practice that drifts from its trajectories in order to track moments of encounter. Rather than develop a representational understanding of the significance of blue in its various contexts and as a container of hidden systemic political and economic operations, I press in on moments of blue contact, impact, and intensities that form as emergent atmospheric pressures. I turn to the eventfulness that erupts out of connections and couplings of singular, virtual, potentializing qualities of blue actualizing worlds through a relay of sensations tracked as generative modalities of seductions, vitalities, attachments, wishes, nightmares, encounters, successes and failures, for better and for worse.

It's not so much about signifying as it is about tracking a moment of intensifying, a singular worlding augmentation. What is significant about blue in Belize is less vital than carefully attending to the ways in which blue materializes, emerges, attunes, and attaches as innumerable linkages and flows of proliferating, everyday village life compositions or expressions of things actualizing. I press in on provocations of blue forces, arbitrary scenes of living that conjure compositions of life in hues and feeling tones of blue intensity, which enable atmospheric pressures that are the sensing modes of living as they come into being: obstinate, struggling, restful, promissory, across real and imaginary social fields, shifting lifestyles, weird circumstances, and states of immersion. Such provocations are posed by forces of material indeterminacy with forces of living bodies by efforts of networks, assemblages, territories, and temporalities that impede indeterminacy enough to extract from it something intensifying, a performative organization of blue hues and feeling tones, a generative coherence of forces, blue as devotion, that seduces and enables life and next steps.

Notes

This chapter has been revised from its original publication in 2015 as "Belize Blues" in *Semiotic Inquiry* 32(1,2,3): 25–46.

1. This poem is deeply inspired by Maggie Nelson's *Bluets* (2009). I wish to acknowledge Nelson's linguistic originality and her insights into the color blue in its creation.

Chapter 7
Parca's Picks

Thirty-three Whoopy, or Numbers with No Relation to Number

Parca says that I met her on the rebound. She was off her numbers, when her life was doubly difficult: hard to follow and hard to bear. Until it wasn't. She took interest in me because I was the *come and go* white guy some locals were talking about who had hung around Wallaceville for far too long to be a tourist but not long enough to be an expat *come-stay* but who started playing Boledo and losing until he began to win playing the number thirty-three, once three times in one week and one of those times in what he thought was a big way, enough to pay two month's rent. Parca was curious and then nosey. As hard as it was for her to do it, she much later admitted, Parca hailed me in the street one day to nervously ask what numbers I was going to play. This was something she would never ask me again even though we talked about picking numbers almost every day we spent together. We still do. I could tell she was serious. Unguarded I told her that I hadn't thought about numbers or playing. *Think about it!* Parca snapped.

Hardly thinking about it, later that day I dropped into the Chinese market to buy my Boledo numbers. I picked thirty-three and won again. Parca did too. She had picked thirty-three as well, and a series of ancillary numbers that thirty-three would surely call upon soon, and did, because of how these numbers *follow* each other, how they *call on each other*, how they seduce each other as they condense a combination of sensations, or how they collectively resonate as a vibrant collection of felt things and how they are numbers that hold "no relation to number" but are instead the stuff of the durative and of a processual multiplicity (Bergson [1910] 1950: 75–139).

Parca said she would share those numbers with me this one time if I wanted them. *Yes, I want them. But how did you know to pick thirty-three?* I asked her. She smiled and whispered *whoopy*. Some Belizeans say that when a number has whoopy it brings on more luck and an intense feeling for stronger numbers generating better powers of prediction. But whoopy isn't just logic and luck, it's something special conjured out of repeated things and wild forces that activate some swirling flurry of desire for Parca. Parca says thirty-three has whoopy. Maybe it's just hers, maybe mine too. Things get complicated with the *maybes*. But one sure thing is that Parca wins big playing the Boledo and with such frequency that *it's crazy scary*, her friend says. *Parca*, Toycie says, *has some pick power. You find imitations of Parca's special cool, but she be new every time.*

The Panama

The Boledo is the oldest popular nationally authorized gambling game in Belize. The word is derived from "bileto," Spanish for ticket, and a similar game called "the Good Old Panama" brought to Belize in the mid-1940s by wartime Belize workers returning from Panama. The Bileez Kriol language gloss for the word Panama is "taking a chance" because that's what it took to leave Belize and travel to Panama during the war and then return a Panama winner, a lucky move, maybe, and to play the Panama, a lucky guess, maybe, because there was no guarantee of payment upon winning at a time when the game was informal and unregulated. Belizean workers returned home with a *double-digit* habit. They arrived *blade up*, *dudes* with fancy rings on their fingers and US greenbacks in their pockets possessing more money than any Canal Zone wartime Belizean laborer ought to have, to take up a propitious flirtation with a new local numbers game. The Panama was illegal until 1945 when the Lotteries Control Committee of Belize was created to administer its Belizean version, the Boledo, and discourage cheating and future feuds between crooked agents and players. Officially the money to be made was in selling numbers not in buying them. But for the buyers and the sellers the game thrived on amulets, dreams, names and charms, flesh, bones, and coastal signs of sea life and land forms, on convulsive moments of fear, panic, luck and the nerves that built up in local bodies as the forces of things piling up in the rushing waves of stories, events, and objects, incessant compositions growing so dense and textured that they became an unearthly liveliness that was actively, if modestly, fashioning this small and unnoticed colony.

Wallaceville was an insular little village in the 1940s while Belize was still a backwater British colony invented and codified in waves of influence by pirates, explorers, trade merchants, religious minorities, loggers, and colonizers, looking Janus-faced to Central American woody interiors and Caribbean marine life and water worlds. Belize was described then as fragmented, unstable, reciprocally isolated within, culturally heterogeneous, lacking historiography, communication channels, and historical continuity. It was described as impermanent, syncretic, a discontinuous conjunction of unstable condensations, Obeah turbulence, mangrove swamp, bugs, jungle, rivers, reefs and ruins, oppressive muggy heat and disease, ineffectual British colonization, pirate holdouts, "grubby" coastal Carib villages, and Maya mission stations. Such is the shape of a materializing Belize chronotope of the Boledo's emergence as the game began to seduce Belizeans. And by the time Belizeans started officially to call each other Belizean, in 1981 when Belize became a nation, the Boledo had found great popularity and a formal support structure. Chinese merchants, immigrants from "Formosa," through loose and competing networks of general market stores, were given state permission to run the lottery and to license local agents who, through early evenings, sat on fence posts or set up makeshift roadside tables six days a week to sell tickets for an official cut of the action, or maybe more.

That's how Parca got her start playing the Boledo. She and Wally, her first *long-time* expat boyfriend, had a license to sell tickets in Wallaceville in the 70s and they did so successfully for several years. But Wally was a drinker. Parca was too. They easily lost track of things, losing money, days, and numbers, not paying out on a win, and then surreptitiously cutting in on the regulated game like secret agents to play an illegal version of the Boledo, and so pay out or up only when they were forcefully reminded by winners to do so. That's when Parca was young and could never seem to win at anything. That's when they had to stop running their game. That's when Parca was stabbed *bad* in the belly by *Bastard Fuck*. That's when Wally became very ill and finally sailed out to one of the cayes to kill himself leaving Parca with heavy debts, in deep trouble, with two small kids to care for. But she managed to climb out of that hole and promised never to run a game again. Parca, however, never stopped playing the Boledo.

A Boledo ticket records the numbers that you pick to play. It becomes the record of the pick, today recorded on the special *Chinee* store, government-certified Boledo computer, and a ticket is printed out on a special *Chinee* store, government-certified Boledo printer. Tourists and expats don't play Boledo, Belizeans do. That doesn't rule the tourist out. *They are useful when they are useful*, Parca explains. *Just like you are Ken*, she smiles. The rules for the Boledo are simple. Pick a number or several numbers from 00 to 99 and place a bet on those numbers. There is an official calculation that determines the ratio of the amount bet to winnings; it's based on the number of pieces purchased. A "five piece" is the smallest ticket you can buy. One piece cost a shilling, BZ 25 cents. The payout is BZ 17.50 if your number plays. Numbers are drawn on weekdays and then during the more expensive Sunday morning Lottery game. Winning numbers are announced across the

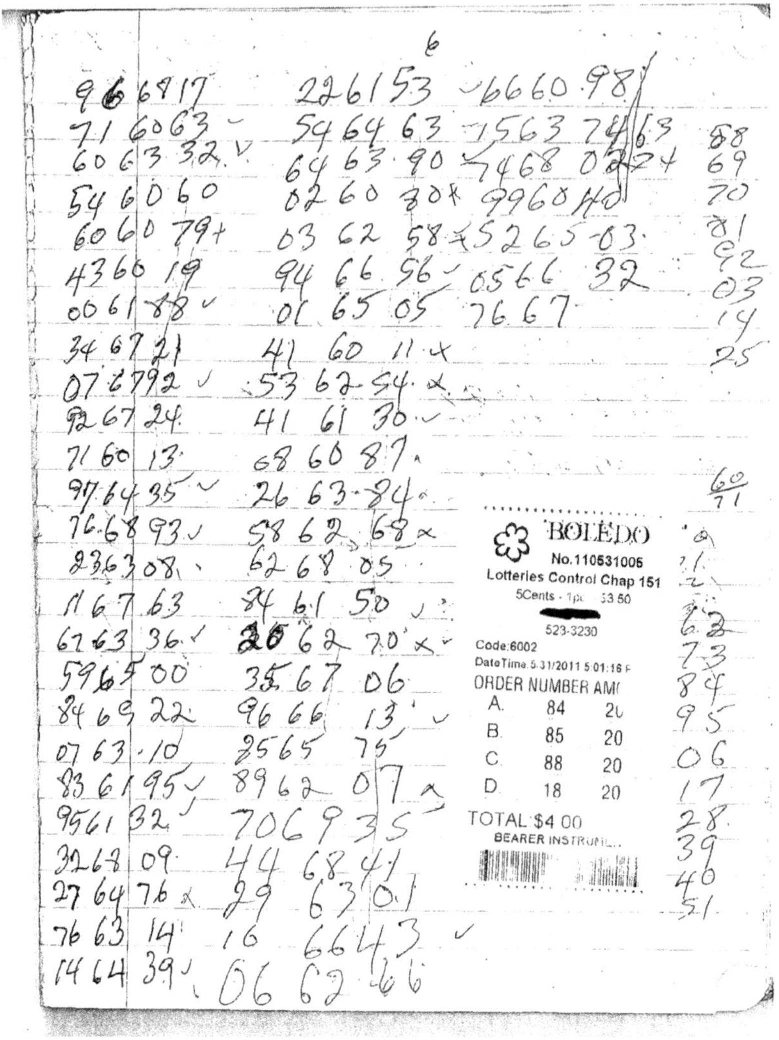

Figure 7.1. Parca's numbers and a Boledo ticket. (Photo by the author.)

country on Love FM at the same time each weeknight, eight o'clock sharp, and then at ten o'clock on Sunday mornings. *They always have been*, Parca says. Bet as many pieces as you feel like you can or need to bet. Put as much money as you like on your picks. But after that, things get complicated, at least if you want to win big and often like Parca does.

Belize changed; Boledo didn't, except for the technologies of the pick. It used to be that a pretty Belikin Beer girl or some reliable national notable would pick numbered table tennis balls from a hand-cranked basket set up in the Biltmore Hotel in Belize City. Live on the radio. They stopped that practice because it was too easy to *pick cooled*. What used to happen, so the apocryphal story goes, is that the winning ping pong ball was injected with water and then frozen so that the picker could find the heavier frozen ball at the bottom of the cage and pick the prearranged number. So the fix was in for those scheming leaders and organizers who were in the loop. The heated stories that accompanied this revelation, never officially confirmed, were legion and flowed like lava, causing so much hot havoc that it fried what foundations of national trust leadership institutions had been trying to establish and exemplify. The more things stayed the same—risk averted vertical hierarchies of national power, privilege and knowledge—the more things stayed the same. Today it is the "safer, more secure" computerized digital random number generator that makes the pick; the use of street agents was dropped a few years ago and replaced by computers. These were set up in the *Chinee* stores. The risk is spread out now and *democratized*. No one person of power has an advantage betting and picking. The real millions are made by those members of the national consortium that run the game. But for the most part they don't play the Boledo. *Why would they play?* Parca asks, *They own it.* But the "they" in Parca's imagination is some secret cabal composed of some *big Chinamans*, foreign economic interests, *maybe Coca Cola*, Parca speculates, the Belize Bank, each attached in its own ways to wild fragments of national Belizean business interests, money flows, drug cartels, NGOs, and maybe even churches who mix with members of governmental power elites. In Belize it means they are pretty much the same people.

The digital mechanics of contemporary computerized Boledo gaming and its networks of control indexes the neoliberal economic "logics" of a decentered state now fashioned through formulations of security, safety, risk calculation, statistical prognosis and probability. So local bodies become understood not just in terms of some representational essentialism as traditional Creole bodies and cultural artifacts of the state but also in terms of informational coding practices broadly attached to Belize as a transnational tourist state (Berardi 2012; see Little 2010). Here bodies are always in the processes of correction and readjustment through multiple agencies—foreign-owned banks, eco-NGOs, Christian churches, expat social projects, international holiday and resort property agencies, and tourism developers—all issued as imperial forces over life and each urging an ongoing refrain of better risk coding (see Piot 2010; Berardi 2012). And here a middle-aged Creole woman is herself variously assessed based on disease, fertility, religious fervor, environmental safety, climate change, the WHO, the World Bank, cultural knowledge, reef management, or through some other management-system-agency logic (see Massumi 2015).

But here, Parca, an aged, classed, gendered, racialized Belizean body, is productive, too. Not lacking the capacities for re-organ-ization, Parca feels or senses how to live in networks of control through the eventfulness of her picking power, in relation to the challenges of chance and risk. Parca's picks press in on the assumptions of expected neoliberal logics of assessment and life formations in this rapidly transforming place, now gone to the tourists, expats, and NGOs, and so picking is a generative practice of liveliness that actively disparages life captured in the given structures of a Caribbean political economy of tourism and frees up the euphoric surge of freedom occasioned by the care she feels for her acts of picking themselves, whether successful or not (see Berardi 2012; but c.f. Rancière 2004; Agamben 1998).

So now, each weekday evening, long lines of locals reach from the front check-out counter to the dark back edges of the old Top Star Market, down aisles of hair and grooming products, baby diapers, and toilet paper that smells of "Spanish" laundry soap mixed with buckets of rotting pig tails. Air thick with smell and sweat and nerves. Everyone anxiously checks their numbers over and over to themselves, clutching soggy paper lists of picks that stick to skin, dissolve in hand, impatiently waiting their turn to proffer them to the *Chinaman* in hushed tones or by anxiously shoveling long lists of numbers at him as he quickly and mechanically enters them into his machine. Everyone watches carefully that he doesn't make careless mistakes. Neighbors watch and whisper to each other in soft, hushed neighborhood-night tones in the sweltering heat of the store, straining to hear or by chance to see another's picks, making last minute adjustments to their lists, or not, the sounds of hushed voices washed out by the racket of an ancient ceiling fan that must keep its own secrets about bad picks and winners in the agitated clickity-clackity rhythms of its wobbly rotations inciting the emergent nervous matter of picking to grow palpable.

Now it is my turn. The *Chinaman* leans in on my breath like a priest hearing my confession, struggling to make out my muttering, second guesses, exhausted hot confusion, and hesitations.

42 fifty pieces
24 fifty pieces
13 twenty pieces
33 one hundred pieces

He enters my list of picks, hands me my ticket, and takes in the bet. We play. We scatter, our Boledo tickets safely tucked away, and rush home to wait for the eight o'clock draw.

The Pick

You would probably be wise to think that Parca's daily Boledo picks are at best lucky guesses if she wins. They are, but not at best. Parca understands why Belizeans like her, now more than ever, buy the Boledo. She considers the economic hardship most Belizeans are currently facing to contribute to the pull toward the possibility of a quick win. Right now, she says, the way the economic situation stands, a game of chance is as good as any income generator for any worker who sees very little light in Belize's turn to a tropical neoliberalism. The long lines of people queuing up to buy Boledo tickets attest to the fact that there may be as much *Fortuna* in picking Boledo numbers as there is in trying to find a good job these days, or cultivating favorable dealings with the government, international property developers, crooked expats, thieving neighbors, condescending NGOs, tourists, and pesky, overly zealous Christian missionary leaders. For Parca playing the Boledo may have to do with all those things, but there is something more to the conditions of endurance that such dealings generate and that gesture toward that which is otherwise to reason and logic.

Most importantly for Parca, her picks are a force of lively social-material composing, a kind of mattering out of sensations, dreams, and the atmospheric pressures of daily events. They must be deeply felt to matter at all and for Parca to start caring about them. The Boledo is Parca's careful focus on the ineffable and the insensible, her attunement to things in their making, a mode of expression that she has cultivated over years of playing the game with each pick carefully charted in her *numbers* books. Together these *numbers* books constitute an active history of picks and misses indexing a dense tangle of events that are the stuff of her history in this place. *These books are my life*, Parca confided one day when she was trying to figure out what to do with the old ones, now rotten and moldy, earmarked and in tatters. From these she recently took days to copy important number sequences and dates into a new book and then burned the old ones, except for the one she gave me to look after and maybe to consult while preparing this chapter.

Many other *born locals* in Wallaceville have developed their own systems for picking Boledo numbers but nobody is as good at picking numbers successfully and consistently as Parca is, and everyone knows it. But Parca insists that when she wins it is because she feels those numbers *all the way down*, into her blood, bones, breath, and flesh while leaning into that swirl of possible connections: human, nonhuman, virtual, and actual. Parca stays attentive, focused and open to her pick system, attuned to its nuances and the swirling stuff of hot little secrets and prickling big things. Bothering agitations. But winning has to do with a lot more than the money she brings in and the loving things she can do for her family and close friends with her

Figure 7.2. One of Parca's number books for 2013. (Photo by the author.)

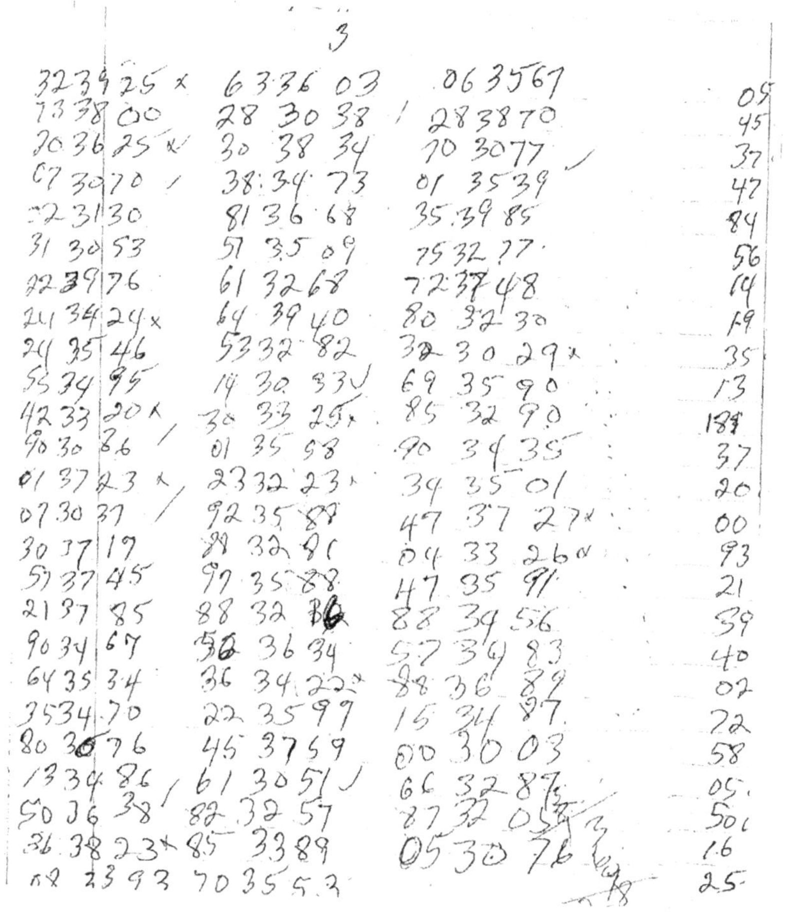

Figure 7.3. A page of numbers from Parca's numbers book, 2013. (Photo by the author.)

winnings, like buying one of her boys a piece of jungle so that he could hack out some commercial farm land, or like buying the other one the truck he needed for hauling stuff. For Parca, attending to her numbers is also attending to life, to her relationship encounters with *everything*, she says. She says I can share her system with others because they'll probably consider her crazy and her system nonsense. But that's alright, she says, and so she dodges prying curiosity in a primitivist trope. She's *vexed*, however, that I am starting to *get it*, that if I learn to focus properly I can win, too. But that means that I may be *getting too close to things*, she worries. It is the care she takes in our

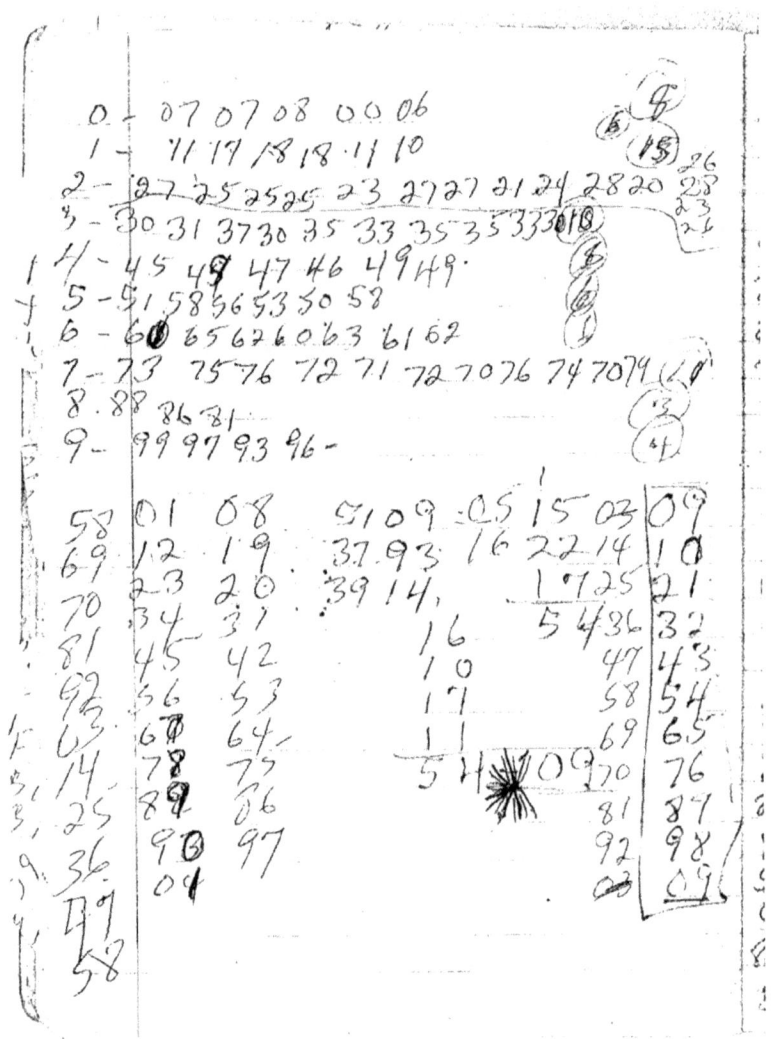

Figure 7.4. One of Parca's number swirls, 2013. (Photo by the author.)

relationship; it means a kind of responsibility that is so deeply felt it scares me. She fears sometimes that she may be teaching me more than I can handle and that I might lose a lot more than my money. *What?* I ask her. But Parca, smiling, won't answer. I should know or at least know better than to ask. But as hard as I try, I am not in Parca's league. Parca wins, I hardly ever do. Except for the time I played thirty-three and attracted Parca's attention in the first place, and so summoned some strange moment of curiosity and a relationship, now long and close, that helped her get back on track with her numbers, I lose more often than I win.

Nevertheless, curious I remain. After the winning spree with thirty-three others also began to ask my pick advice and seduce me with their talk of picks into a lot more than talk about picks. This was talk that flowed into gossip or hard personal talk or confessions about things, about odd moments and relationships that were conjured through the Boledo. Soon others wanted to know what I was talking about when I was talking about numbers. Still others began to feel slighted that we were not talking, that I had Parca's secrets caught in the grasp of picking details that I kept to myself as a selfish gesture or maybe a mocking joke. Surely I had some number advice? What was my secret, what were my numbers? All of this activated a liveliness of its own that started to become trouble the more absorbed I became in the social and material vitalities of other's lives through number picking. Soon there were those who blamed me for their bad picks and lost bets and who then started bad-mouthing me making life a series of ducks and dodges, run-ins and confrontations the further I was seduced into conversations about secrets, lives, and numbers. This is what Parca was worried about, or at least partly. Like Parca, I learned to keep my mouth shut and my picks to myself.

Flash

And that's why Parca wishes I would stick to the rake, the money payout on a Boledo win, even though she feels that I know how the Boledo *isn't just about the money*. For Parca the rake is composed of a lot more than her money winnings. Winning is a lot more complex. When she is in a teaching mood and focused, life feels a bit less than burdened. That's when, as Richie says, that Parca can *pick her numbers out of nonsense*. That's when picking winning Boledo numbers is like a *flash of lightning*, Parca explains. She can smell the numbers like ozone. She can feel their energy, their heat, and intensity, the hard and soft push or pull of their relational pressure on each other and through her. Even with her eyes closed the world lights up in a super-charged flash of number combos that burn Parca's retinas from the inside out. That's when the atmospheric pressure changes, the hair rises on the nape of her neck, her arms and legs. Her belly boils. She becomes charged, a transformer. Her scalp turns into a *wild itch*. Sweat wet she can easily pass out. Turned into a conduit of transitional contacts caught up in flesh touching bone, touching breath, touching tides, atmosphere, time and space, it is necessary for her to stay with the intensity of this act of emergence or the flash will push in on Parca like the pressure of the world or like some corporeal immensity, pressing in on her, or like some invisible atmospheric pressure that envelops her like the clammy wet weight of an overcast day.

That's when things happen, where things change, Parca says, *in the flash*. Out of the chaos of the flash, things emerge and compose themselves in these affect-laden entanglements of potentially transformative social-material encounters. For Parca this is all about how to get *inside* some things and how to learn to *leave other things alone* in this new world of composing entanglements, an emergent social real in Wallaceville, a tourist town, in Belize, a tourist state. Then her dreams resettle and her body calms for now, until the next pick the next time, that's if she's lucky.

Parca's flash is a contact zone of roiling multiplicities that figure through an unstable yet forceful and urgent congeries of sensations to which, on *the good days*, she is receptive and open. Her picks are acts of creative variation hailed by things happening that stand out for her in a way that she can feel, but they are contingent instantiations of her entangled relationship with the undifferentiated vitalities and energies of her ancestors too, through which Parca generates the relational resonance that numbers, when she is attuned, can *bring on*. Parca speaks insistently about the uncertain gap between numbers that she feels in sensations, like tingling shivers or indiscriminate and sudden little pains, tugs, pushes, pulls, smells, acid tastes, the pressures of tension and release, that she says spread through her body in waves that ebb and flow. They *nag* her. She calls them *visits*. They don't stay long but they won't leave her alone either. They exhort her to respond. They manifest in bits of matter, in everyday events, sensations, and dreams that continuously compose and decompose things for her. Such things *glow*, Parca says. Parca's picks are generated in this transitional gap of an emergent relationship and in the things that stand out because they *glow*. She feels the attraction of the luminescence. A number or a set of resonant numbers unfold, illuminate, before they fold back into some puzzling proliferation of energies that will again unfold through another variation in stuff and dreams, sensations, everyday events, as swirlings that repeat themselves, each repetition an ongoing difference.

Parca's picks, as fieldings, as swirling vibratory relationality, are intuited, but Parca's intuition is not some mystical state that she enters on a whim. It is not an option for Parca to be so intensely involved in such a movement of "thought-feeling." "Intuition does not have an object," Massumi (2015: 45) says, "It has a fielding. It comes with a field potential that is movingly thought-felt before its elements are consciously registered as the object of a fully formed perception, and is immanent to what occurs as a function of that field." Parca cannot linger in the dynamic gap of an incorporeal materiality that is the transitional fielding or swirling of her flashes, what Deleuze and Guattari (1986a: 157) call the field of immanence. Parca feels in transition, in variation, when in the act of picking out of the virtual dreamworlds she is in touch with, as such touches on a world of feelings, matter, and events. Fielding a pick for Parca is realized retrospectively, a number is conjured out of particular things and situations and identified feelings, but each pick figures only because there is an infinite resource of other feelings, situations, and things to figure with (Massumi 2015: 45). When things are going well Parca feels numbers resonate and so create this transitional space of opening pathways. Even if those paths are still indeterminate, they are mutually included in the event of the pick. Picks are moves but they are never the same move twice. They are variations that mutually coexist. Hardly resolutions, Parca's picks, whether successful or not, are intuited and such intuition is the stuff of her attending and attuning, her swirlings (see Bergson [1910] 1950: 95–97).

The Fire

Thick smoke woke the village from its middle of the night slumber. Coughing and choking, I thought my place was on fire when the reach of Richie's voice had me bolt up and out of bed. Still uncertain about things, I looked out over my front deck and down at Richie on the beach. Bright fiery embers were flying into the dark sky. They silhouetted Richie in a crazy, menacing way. Everything was confused and nothing smelled right. The scene behind him seemed to have been lifted from *Hell*, the third panel of Hieronymus Bosch's *Garden of Earthly Delights*. It looked like the whole village was on fire. *Haul your ass*, Richie yelled, *Ruby's restaurant is on fire!* Ruby is Parca's sister, Richie's cousin. We ran to join her bucket brigade to find that three other places in the neighborhood were on fire, too.

Pail, hand, water, hand, pail, water. Next one, next one. There is no other way. Don't stop. No time to chat. Focus. Faster! The volunteer fire department finally showed up and with an impressive speed and efficiency hooked up its hose and turned it on. The hose dribbled water as if to intentionally frustrate the rescue effort and encouraged a panic proliferation of conspiracy theories and ridicule of the emergency itself. The water tank on the back of the firetruck. Empty! Someone forgot to fill the tank. It's a Keystone Cops slapstick situation tragedy (Berlant 2011: 6) and the whole brigade can't help but laugh sadly, disgust assembling with exhausted derision, bodies too weary to be angry. Fear now. Tired but deadly concern. And bodies in the grip of what Parca calls *the dreadfuls*.

Peals of laughter mix with voices of disbelief, muscle effort, drenched bodies struggling with the collective effort to save buildings. The roar of the fire is deafening and frightening. Wood walls and thatched roofs crackle with a shocking mockery and finally collapse into huge plumes of dangerous burning *chit* that reach far into the sky and float menacingly through the village and over the water on a stiff night breeze. Madness and threat mix with horror. Cloying smoke steals the oxygen so searing lungs collapse. Embers burn scalp and skin. There is the smell of burning hair. Fatigue quickly sets pain. We are all unnerved, so tired and terrified. It's a struggle and a lost cause. The building we were working on burned to the ground. Our effort turns to saving the homes connected to it. It is the same for the other fires; all but one of the homes burned to the ground.

At that moment I looked up from my brigade work and into Parca's eyes. She was frantically looking for Richie. She needed someone to look after Ruby and her mother. Both have now lost everything, the restaurant, their means of employment, the business they loved and worked hard to improve, home, clothing, property: everything. They are both inconsolable, in shock. Richie leads them away. Parca stays to help, but it is all too much for her. She collapses in the dirty sand, in shock, and yet when she thinks about what just happened it is a matter of trying to settle deeply frayed nerves, anger and confusion set against the big question about how this happened. Later Parca said that it was a crushing experience: the scorching fire, sickly suffocating smoke, blazing embers, a searing breeze touching her skin and soul and it all fed a furious rage and incendiary emotions. Burdened by monstrous fire, she felt half-crazy. But then the numbness set in.

For weeks the village smelled charred, plumes of smoke still appeared here and there at the fire sites. Shock still seized and squeezed nerves, an impossible nightmare slowly actualizing. We were told by the Belize City inspector a few days later that the fires were intentionally set. Structures burned easily and quickly because each of them was built with old wood and thatch. That's how the arsonists picked the places they did, they were made of easy to burn dry thatch; thatch, the very material of tourist seduction meant to create that tropical happy vibe that was a "go to" index of Paradise happiness. Soon anger and panic set into the nervous system of the village and it touched everyone. People argued with each other about who was sick enough to do this? Why? Local tourist businesses just getting off the ground; who had the most to gain in destroying them? Resentment flashed with every possible guess, the winners and losers, saints and sinners. Fingers pointed in many different directions resurrecting old grudges, unpaid debts, unfair competition, ethnicity, class, the future of the village, local corruption, politics, big conspiracies, bad land deals, idiotic moments of social ugliness, jealousy, and violence—it all served to further unsettle sleepless souls.

Mr. G. said there should be a lynching soon. That didn't happen. But there was a collective sense of things coming undone and spinning out of control. Everyone was moved, Parca especially. This was *personal* and *close* for her. Parca and her family were touched deeply by the fire and responded to it constrained by the usual political terms of village participation. Like others, once the shock cleared traumatized bodies, they became touched by a fiery anger and behaved accordingly, joining the social circuits of rumor and gossip, accusation and allegation that were the lively energy of an angry, frightened, and growing, village chorus, the energetics of public secrets defacing local life itself (Taussig 1999).

I could see that Parca was also trying to work with the stories of fear, suspicion, and accusation that animated daily living after the fire. A sooty tactility, a visceral attunement, animated by the hard and soft soiled surfaces of charred matter and alarmed voices, pulled Parca into some precarious alignment with things to produce some social-material emergence of threat. All of this seeped into Parca's heart, skin, dreams, and bones, and her body took in the smell, taste, and touch of exhaustion, panic, resignation, the changing light of the village, the smoldering reconfigured scenes of sand, sea, and charred coconut trees against a new view of the tourist sidewalk. Parca attended to her family, her village life recomposing in new circuits of intimacy and care. Her father began to appear to her in her dreams, as real as anything else she could feel.

do you remember?
he asked
do you remember how to get free?
how to build your Creole freedom muscles?
he asked
how to protect yourself,
move through your own cycles?
he asked
what do your instincts
tell you father?
she asked
ancestors prepare dreams and clues for you
they're trying to figure things too.
he replied
can we figure things together?
she asked.

Dreams of her father played on one theme-clue-thought-feeling-fielding, that tragic moment some years back when old Fred had to sell a big piece of beachfront property to pay for his cancer care. He sold it to a rich Lebanese-Belizean family known regionally to take advantage of situations like this and for its brutal disregard for local social life and for its ruthless business practices. It was a horrible moment and a very bad land deal made out of necessity and chance by Parca's father. He died soon after. But after medical travel, and funeral expenses, Parca's family still had a bit of land money left. Truth, pain, endurance, and a hard lesson learned about holding on even in the worst circumstances, they put the money into building Ruby's restaurant on family land. It fast became a very popular bar and restaurant right on the tourist sidewalk attracting locals and tourists alike. And so it looked like there may have been a silver lining to all of this yet. The restaurant's popularity and success, however, did not sit well with the Lebanese family who had developed their beachfront property into a huge nightclub and an outdoor restaurant bar of their own.

Some say that they actually heard the head of the Lebanese family grumbling about Ruby's place, that it was *in the way* and that her family *was nothing but trouble*. Others say that he was known to fight with the village council and completely disregard zoning laws, licensing and building restrictions, and noise regulations in his fight for the largest share of tourist dollars in the village. And he was supposed to be very good at paying people off. Ruby couldn't afford to fight with him, and she wouldn't be bought off when he wanted to buy her out, and so she ignored him, the gossip, and the bad-mouthing *rass*. Still Ruby's place prospered and grew. After the fire, the sour flare-up of rumor mixed with hints of innuendo flew through local air like burning leaf the night of the fire and it landed hard on the Lebanese family, lighting its own political flames. But no one could prove anything. There was no case to be made, but everyone still to this day has a feeling.

Shortly after the fire Parca woke from another dream about her father to the sounds of his voice. But that was impossible. She was unnerved by this until she was able to figure out that the voice was actually that of her father's brother who, along with her brothers and a few friends, was working on clearing the fire site. They were calling to each other, deep into an intimate age-old rhythmic banter, dripping some sweetness of bodies wrapped in laughter, hope, and family solidarity. Parca swore that she could hear her father's voice in that verbal play. It was the tone, timing, phrasings, and accents, the force of cadence, its intensity catching something of local laughter's potentiality drifting in and out of touch on a light beach breeze that caught her up as an alluring dreamy seduction and a deep longing. The voice, so deeply recognizable to Parca that she ached hearing it and then smiled when she recognized the ribald teasing rhymes her father was so good at performing. Here something imagined collected, etched into saliva and breath and waves of sound. Voices set on each other in laughter and joy even as these rhymes were also deeply hurtful historical speech acts, rhythms that carried the unspoken burdens of colonial servitude and trauma of inland tree cutters generations old, rhymes centuries in the making. Here, in the funny bone tactility of these historically ambivalent ordinary sounds, tiny acts of emergent social intimacy were setting things in motion for Parca.

Parca attuned herself to the reactivation of the desiring force of these bodies: voice, touch, sound, producing new conjunctions of flesh and meaning, rhizomatic connections irreducible to the operating functions of language. Voice (not language) as sensation, sound acting, reverberant and wave-like through the body as its singular enunciative process, meaning's emergence in the parataxical piling up of voices, rhythms, sounds, and moves in excess, cacophony, and dream images too, an excess of sensuousness exploding through the circuitries of sound into language loops into experience and the new signs of connection assembling as another of Parca's picks.

It was when she was working through this noise-voice-body-movement assembling, attending to it, to waves of sound touching bone, touching breath, touching sand, soot, and sea, and her ancestors instantiated in her dreams of her father, and the intensity of her father's strong voice, that she began to feel the swirling passionate pull of things composing, a tingling glow, a flash. She felt its historical constraints and those of the fire, but Parca could not help but begin to move with the eventfulness of the situation when it began to become part of some novel constellation of her bodily sensations immanent to the event, encouraging a new social-material fielding. That's when Parca began to consult her dream book again, along with the Grinning Doukie, fielding signs and shaping figures into her numbers book. In short, that's when she began to attend to the relational ecology of activating passions, urgent and intimate, that fashion her through her numbers. And then, once again, Parca began to work with her numbers to set the scene for fielding another pick performance. She felt the fire as event, as a growth of bodily sensations, an emergence that connected ancestors, family, the vagaries of everyday ordinaries, routines, oppressions, dreams, histories, intimacies, anxieties, and in the unsignifying intensities of responsibility when her numbers called her into felt ways of going on in the world. Whoopy.

Bone Flesh Dream Flash

do not dare tear my flesh
from my bones
deny my spells cast in whispers
bury my wisdom with your name, she says

you once slashed at my flesh
and claimed my talent evil.
do not dare deny the
parts of parts that make me whole, she says

> *yes she believes in magic*
> *conjuring with the bones and tones*
> *of those whose dreams she bears*
>
> *she picks in surges of marvel*
> *numbers calling magical numbers*

Parca says her picks are shaped with clues her Creole ancestors drop into her dreams. She is thankful for them because these clues are an unforgetting (after Stewart 1996: 67–89), a process of creative unfurling by which I mean the bodies of her ancestors touch her and in the touch the present world becomes a composing figure. Figures compose, they unfold with an aliveness if Parca allows herself to be lured by her curiosity, wonder, and the mystery that a touch might convoke as an intuition. Parca's dreams fall on her flesh in affective intensities that index her acts of picking numbers. *Follow the feelings that fall into dreams*, she tells me, *They lead you places if you can bear their touch.* Ancestors come in dream flashes: as tingling hunches, shivers, body glitches like sudden stomachache, headache, heartache, bone-ache and shifting pressures that rattle her body. When she is ready Parca attends to such sensations and consults two texts, her King Tut *Dream Book*, when she thinks her dream-themes are getting complicated, and her "Grinning Doukie Skeleton," a map of joints and bones attached to numbers. She carries both with her everywhere, although by now she knows almost everything they have to say, *by the heart*.

The dream book connects dream images with numbers. Clues to picks are connected to breaks, aches, itches, throbs in bone, muscle, sinew and joint that she may feel. She can plot as much with the Grinning Doukie, and attach such emplotments to her dream images and the associated sensations that connect to her dream book consultations. But there is more there (Thrift 2004), and the more has to do with how she is able to make a creation story out of such connections. A story in the sense of sounds, matter, and feelings that enact a "doing" by which I mean a creation story as a compositional practice, a passionate worlding out of the sediments of image and matter, the senses redistributed by the turbulence of imaginative tellings and confabulations, all through her numbers (Stewart 1996, 2007, 2008; McLean 2009). The more is enlivened by the way Parca scans for parallel sensations absorbed in off-shore fantasies and local stories of everyday life, in the way her imagination reaches through her entire sensorium and history of sensations as a manner of inhabiting her body. Someone Parca knows has an accident and hits her head. The rake could be the age of the woman connected to the designated number based on the Doukie skeleton. Parca is then compelled to recollect her dreams and the event together with the multitude

Parca's Picks

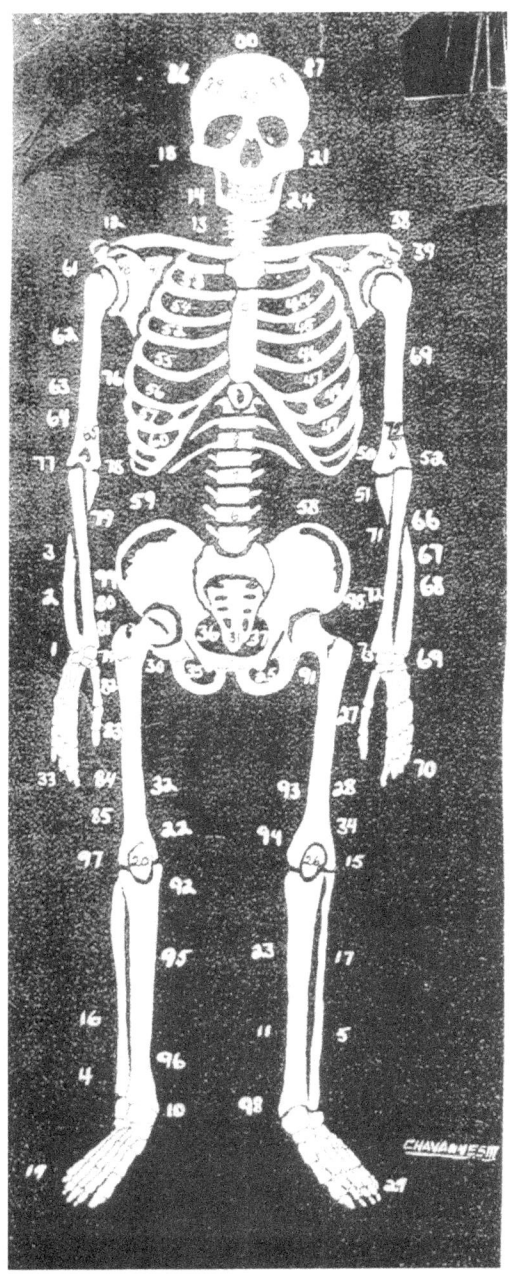

Figure 7.5. The "Grinning Doukie Skeleton," Norris Hall 1974. (Belize National Archives, 2013.)

of other signs from which numbers begin to appear. That is when she must consult her numbers book as things begin to swirl.

These are the shifting elements of a numerical expression. Picking numbers is an activism that thrives, breathes, and sprouts chaotic in Parca's scanning, sensing, and conjuring practices. Picking emerges in the complexity of commingling moments of experience, its unfolding passion, activating a world and its numbers (see Pandian 2012). Every element is an unfathomable multitude because it teeters in the in-between, "on the edge of the infinitely fine blade between being and non-being" (Barad 2012: 210). Barad calls them "virtual particles . . . *quantized indeterminacies-in-action*" (2012: 210). Picking does not prefigure a relational field in advance of the act, rather it's an act of reaching toward through which a relational responsibility might be crafted in a manner "that would exceed the sum of the event's parts" (Manning 2011: 44).

Event

It is out of all of this that Parca feels alive or transitional or in the midst of a transformational movement of negative and positive passions, a provisional harmony of sorts that gathers in the flesh as "elementary particles" (Barad 2012; c.f. Houellebecq 2000). Numbers unsettle things as a power that is no longer indexed to dead biopolitical framings based on law, regulation, or lack and that index the crazy assortments of everyday conditions and events that block and bring her down or make her happy, for that matter. Moving, transitioning, Parca becomes unhinged from a model of meaning and matter based on what Deleuze calls the "tripartite propositional straight-jacket of designation, manifestation and signification" (Massumi 2002a: 21). Instead, her picks are an otherwise quality of becoming, a singularity, a state of things that is not a member of a class of things or an instance of a type of things. Instead, picking activates "relations beyond the linear model of a localized interactivity" (Manning 2011: 45). It is a much more complicated activity than "gridding" (Massumi 2002a: 8). It is the swirling of Parca's intuition, her fielding.

As a swirling actualization of matter, dream, and event, picking numbers is not free of constraint. Parca follows her system of thought-feeling–fielding. She has a method, but it is a matter of following it toward its unfolding, toward the potential that the long labor of her picking unleashes as a "complex relational field in co-composition," a field of relation foregrounding some "*more-than* of experience in [its] unfolding" (Manning 2011: 45). For Parca, picking Boledo numbers is caught up in her attending to things, to an enduring presence as a predictive agency, an abstract calculation in felt movement, when some association of things jumps into a relation as an affirming bodily expression even if it remains transitory, unsteady, or as Parca puts it more adventurously *when things feel shifty wet and unsteady.*

Picking means attending to an anticipatory energy as it transitions wildly between uncertainty and expectation. Picking emanates as a fielding in a matrix of crazy feelings, dreams, and everyday events of which her flashes are composed. *Wild numbers*, as Parca calls them, become regulated in associations and attributions that radiate out of twisting turning lines of force as a deformative composing, as a swirling world-making effort, a space-time of trajectories, lines of incomplete capture that still actualize as something, a mood, an atmosphere, charged up sensations and passions leading to a pick. These numbers are singular and picking them is its own happening, its own event. She can sense the qualitative resonance of a number's relational field in its emergence. Her picks are fashioned out of a particular logic of relations and responsibility, not a logic of measurement or of systemic calculation.

The Visa Interview

When all this works for Parca it works through her, augmenting things in swirls of corelating. But sometimes it all falls apart. Parca thought she had it but then she didn't, a visitor's visa for travel to the United States so that she could accompany Skip, her *next one* expat boyfriend of several years, to the US the next time he took his annual trip *home* for a visit. Skip is a retired GM parts distributor from upstate Vermont, a damaged Vietnam vet whose lingering PTSD dances along with a fiery temper fed by drug and alcohol problems. He can be explosive violent one minute and the kindest kind of man the next, and Parca, as good as she is at figuring guesses, can never figure out who she will meet at the door when she gets home from work. Parca sometimes laughs about being able to pick Boledo numbers but not boyfriends. First it was Wally and now Skip. She sees the pattern, but she hates being lonely and she loves where she lives right now with Skip. It's a very nice place. And while he is detested throughout the village, at least locals say that he *gets it*. He gets what local Creole life there is left to *get* now that the new richer more demanding *condo creatures* have moved into town. If things are bad with Skip Parca will stay with her mother or with one of her sisters or kids for a time. But things aren't always great with them either, as they try to compose their lives working the edges of reckless dreams and hard lives, generating potentiality in endurance, in the fleeting connectives of the tourism industry these days, in their convulsive forces of relation: Paradise power plays fed by life itself in a constant movement of local crisis that passes as normal in Wallaceville right now (see Little 2010).

Skip was preparing for his trip and Parca really wanted to join him since the day Skip had suggested it. Each time he left the village Parca felt lonely and just overall *awful*. She would hide away in his place until he returned, leaving only for work and groceries. Not even Boledo would get her out of the house. But this time it would be different. She easily saved enough from her Boledo winnings for the cost of a plane ticket, the cost of the visa, and some spending money. It took years for her to get a Belize passport, but now that she had it Parca dedicated herself to getting the US visitor's visa. She worked long and hard prepping for the standard visa interview with US consulate agents. It was hard work. First it was the complicated forms and the documents, the endless bureaucratic accountability that she understood very little of, but she dealt with it all. Then it was the interview. Parca memorized facts: addresses, phone numbers, family names and relationships. *How long do you plan to be away? Dates? Relationship status? Just good friends*, she was tutored to say, although they had been living together on and off for the past seven years. Good friends because if said truthfully Parca might be considered suspect, as someone trying to sneak her way into the promised land as a common law partner.

Parca never understood this. As unhappy as she was with the local hardships of her life, its humiliations, physical pain, economic privation, a vexed nation of highly competitive and cheating political elites, ecological breakdown, the garbage, the rich expat attitudes with big city expectations for this little place, and as unsustainable as life may be in Wallaceville, she never felt the desire to leave her family, her numbers, or her life behind for what she took to be the diminished biopolitical framings of life in the US. She heard the horror stories told by local *survivors* of the move, like *Reno, Washington, Miami*, and *NYC*, and by others whose nicknames indexed the places they lived in the United States and whose *American* lives generated an edgy intensity about the place, about keeping bodies alive to work dead-end jobs where success story dreams died; racism, disrespect, poverty, violence, and a lack of generosity and acknowledgement added another layering to a local history of endurance made palpable now as yet another experience of *servitude. Who needs more of that?* Parca asked. That her US agent thought as much worried her even more about whether or not she would feel comfortable there.

Parca had her story down pat, sort of. She practiced for weeks. I filled in as a visa agent working my beat at the US Embassy in Belmopan. I grilled her with questions trying to act like some humorless, hard-assed immigration officer that I could only imagine. We worked on her comportment, her smile. Speak confidently and clearly, *like an American*, Skip pleaded. We worked on her manners on how to sit and shake hands, keep eye contact, what to wear, until Parca was ready. Her story had to be simple, the *truthful* truth. She was encouraged to stick to the facts. Don't embellish. Sit still. Look directly at the agent. Eye contact! Parca was finally ready, maybe, but still she had *a bad case of the nerves* that she could never entirely calm on the day of her interview.

Parca told me later that the only thing that made her feel more relaxed, as though she had a chance of passing her interview and receiving a visa, was if she thought of the interview process in the same way that she thought about picking her Boledo numbers. That was something I wasn't sure of at all and my imagination went wild thinking about how Parca might blow the interview with answers to easy questions using protracted stories of numbers, ancestors, dreams, and making crazy social and mystical Obeah connections between the spirits and US law. She would be on her own with the embassy agent officials and the entire weight of the US on her shoulders pressing at her story with no one or nothing much to help her. What if she started to panic, what could she do? I told her to imagine the agents in their underwear, but that just confused things more.

With two days to go until the interview Parca began having crazy dreams that made no sense but ran through themes of ruin, invaders, and ancestors:

> *daddy was sleeping by me on the night of the hurricane*
> *he caught me with my hand outstretched*
> *when the wind blew the house away*
> *think of who saw us*
> *our father who cannot be in heaven*
> *feel me falling*
> *october 8, 2001*
> *feel 10, 08, 2/0/0/1/, 20/00/01/21 . . .*

> *i can feel these strange men*
> *meeting for the first time*
> *minds as fast as geckos*
> *agents like thick clouds of blowflies enveloping me*
> *senses saturated, body invaders, nonsense*
> *my numbers, I need my nu8mbers*

> *too many of them*
> *i fall from earth into sea: drowning now*
> *too many of them*
> *i need him. he's not here for the first time*

Parca's words tumble now in a mad rush to get it all out, to tell us what happened in there, with those agents, with that one agent. In Skip's truck now, in the parking lot, in the shadow of that massive, white, overly supervised US Consulate, she can't speak. Skip, six or seven beers in, wants to hit her. Still mute and in shock now, Parca flinches, hands up as if to block the blow. Ruby and I try to stop Skip with our voices, but he beats us back with sound and spit:

> *Why couldn't you stick to the script?*
> *... How hard is it?*
> *... Stupid cow!*
> *... A wasted day.*
> *You're paying me back ...*
> *Do you hear me? Every fucking cent!*

We all feel the moment, its horrid growth into some monstrous confusion. The air is thick with anger and fear and angry air energizes the atmosphere, makes it electric and we all feel tiny pulses of crackling chaos.

Parca panicked. Parca told the agent that she and Skip were married, but she couldn't produce the marriage license. Then she said that they lived together common law, a kind of marriage. A crooked story in the making. Then she told *the simple truth*. She told the agent how much she loved Skip and just wanted to go to Vermont for a visit and return home within a month. She didn't want to live in the United States.

US Immigration: *Where's home?*

Parca: *Why, with Skip.*
No, I mean I live with my mother, in Wallaceville.
I mean I have a full-time job.
I ...

It may have been the shortest interview ever. In and out in less than a few minutes but then a long wait in the freezing cold air-conditioned waiting room for the official answer. And then all Parca could remember was that her body was slowly melting while walking across the blinding hot consulate parking lot to us in the truck. Parca said she felt stuck the moment she entered the room, she couldn't catch her breath, she wanted to make a fast exit. Her head hurt, her stomach too. She felt like she was going to be sick. She couldn't think. Interference. Bile. No grip, yet all stick. Sinking now. Nosebleed, stomach pain. *She feel the sick of that*, Erline said later, *like if the tide pull right through you. Whossssssh! All over. Nothin' left but sticky sick*. Something undoes her, unleashing impulses of affliction that pass through Parca, that pass through all of us in the truck, an affliction bordering on a collective delirium of disappointment. All bets were off. We all got dangerously drunk driving home. And after that Parca was stuck.

Sticky Numbers

Picking numbers is a material mattering. It is not a disembodied or uniquely human activity. Stepping into pick possibilities, straying, swerving out of bounds, diverging, touching down again, driving off the beaten path, returning, not as successive moves but as experimental challenges of expression as forces of entanglement, difference, the sensations of entangled things and beings becoming together, becoming apart. Don't lose the thread. Repeat. Feel. Sense. Difference. After the visa scene, there was no thread, no touch, just numbness, no assembling, no movement just smelly sticky sick stillness.

In moments like these Parca loses her pick power. She can't *pick right*. She becomes less attentive to things happening, less attuned to them, more distracted, and the creative drift of her attending and attuning to numbers gets stuck and she loses touch of the dizzying feelings of things transitioning, resonating, swirling. When Parca's consular appointment went rogue she went numb. Numb, Parca couldn't follow through on a shiver, a dream theme, a relationship, a situation that might need her close attention, some undetected isolated event, the call of some bit of matter, some pressing pressure, release, the weather, an iguana's appearance. Her numbers no longer added to things, let alone added up. Not these separately but any attempt to fashion some collective compositional force confused her and morphs into some mixed-up feeling that simply exhausted her.

That's when she says *life gets down* on her and so she gets down on herself. Things start to weigh heavy, numbers *feel thick and hard* and she can't *move with them*. That's when numbers *stick*. Sticky numbers are hard to work with. She can't extricate herself from their grip to work with them. There is more pressure. The pressure builds. It is relentless. That's when Parca will start to play the same numbers repeatedly, regardless, with big bets, and that's not good. Panic and fear can set in and she may have a few more happy hour drinks than she ought to, get into a fight with her mother or with one of her kids or with Skip. If that starts and she doesn't stop, she completely loses touch with things. When she loses touch she doesn't attend to things, she gets caught in centripetal forces and feel like the weight of things is all hers, pushing her down, condensing life to a dull slog. It's a matter of lost balance. That's how she felt after the visa debacle.

Add to this, all the rest. Parca works long hours as a staff cook at a super high-end tourist resort, standing over hot stoves in blistering heat, cutting, chopping, cooking, and listening to the other workers complain about it all. It's ruthlessly exploitative, cheap, and constantly monitored labor that demands an energy and stamina that Parca finds increasingly difficult at her age. Her legs ache. They swell. Her feet swell to twice their size and hurt so badly that she teeters and sometimes collapses. She gets chest pains and bad headaches. Her stomach acts up all the time. Her stomach has always been *disorderly*, ever since she had part of it cut out when she was a teenager, she says. *Now I have a plastic stomach*, she complains. Whatever that is, it touches on pain, the memory of long suffering, village violence, poverty, exploitation, and lots of local death, an associational assemblage of connections, divergences, congealing experience, and foggy, diffused lines of impacts. And when she is unlucky enough to miss a ride into town, she limps the long journey home from the resort to Skip's place, a walk that often feels like it takes all night.

Bad picks are contagious, too. This time they started with her failed visa visit. But they could have started at work, a bad word about the way Parca won't share her numbers, or a slight that lingers and festers and ramifies through village gossip that actually hurts her, and always distracts her. Or, it may be Pastor G's scary, spitting vitriolic of evangelical fervor, *in the name of Jesus our Savior*, thrown in her face by some newly converted local telling her that she is damned unless she changes her superstitious ways, her lying, her drinking, her gambling. Or, it could have started with Skip's impulsive *rass*, another source of exhaustion.

That's when Parca just can't take it any longer and so she *closes down*. But with nowhere to go she endures the pain and disquiet of despair. That's when *things just overload me and vex me bad*, she complains. That's when she feels *jammed, backed-up, hard*. That's when the flow of expression, the vitality of her picks, stagnates and all she can feel is the dead-end of things and the sickly stickiness of her life. That's when the transitory compositional force, or the potentializing passions of the pick is captured and thickens in her misery and congeals like a rock and a hard place. So if it's not some physical hardship interfering with things it's some miserable sorrow that pops up, or some psychic anguish or, more likely, some horrid mix of these things that mixes with the general biopolitical framing of life in this tourist place that makes Parca feel the rough edges of hopelessness. These things put the grip on singularity as an instantiation of the chance occurrences of Parca's picks. It happens all the time, this move from the singularity of emergence through the labile forces of her numbers into the sticky goo of an unsustainable biopolitics of life, from ontogenesis to newly vectored forms of social reproduction, non-duration, stagnation. Stagnation spreads and Parca feels the life drain out of her numbers, and so drains the vitality out of her. When that happens I can usually find her holed up at home or in a local bar, depressed or worse, hollow and hard, waiting it out.

But for Parca nothing tingles with chance, until it does and, if she is lucky enough to shake things off and feel in touch again. That's when she says she can become *unstuck* by something or someone that comes along, a diversion, a curious tingle, smell, noise, taste or some sensation of becoming, some atypical expression of energy that begins to matter. That's what happened when I came along and began to win at Boledo with the number thirty-three. Good things or bad things, such things enter her dreams and begin to agitate in her body. Sensations, events, and things call her, and she starts to prepare again, yet differently this time. She attends to this difference, cultivates it with hope and care. And when she is lucky, she can rebound. Whoopy! It happens when it happens or it doesn't. It's always two steps forward, four steps back.

Response-ability

In times like these Parca feels the burdens of responsibility press in on her *like the grip of giant hands* and picking numbers *twists me up*, she says. She says this should concern me too; and it does, deeply. I feel a responsibility too, to Parca, to my writing through Parca's Boledo picks, to envisioning some otherwise liveliness transiting as an emergent sociality like a fictional truancy of fantastic stories of demons, fires, visa agents, nightmares and dreamworlds, bad trips with libertarian pretend pirate expats all twisted yet becoming a consistency of experiences, a "multiplicity of immanence" (Deleuze and Guattari 1986a: 156–57). Parca did what she usually does with such events, she began to trace them out in her numbers book. Parca's books of number picks are her *experiences*, but each next pick torques the storyline of experience and offers a twist on the frameworks her books offer her, they take her out of frame yet she stays in touch. This is not a book of numbers as an "unfolding algorithm" (Barad 2012: 207). Parca's number books are not so much stories about the everchanging world of Wallaceville, not so much a resistance or even the resilience of a life lived there, instead they instantiate collaborative composings as counter-actualizations, books of numbers that are generative: resources, yes, but possible incitements too. Here ancestors, coral reefs, her sister Ruby, iguanas, blue, bad-ass boyfriends, jellyfish, her mother or her kids, grey atmospheres, red snapper, messed up work life, daily pains, sand, rain, sun, soot, the cayes, stray from calculable numerical reference points to become forces of expression potentializing in touch, in the touches she leans in on.

This is how Parca conjures her picks too. It is a matter of "collaboration," what Barad glosses as "to be in touch in ways that enable response-ability" (2012: 208). Not as a sign system but as "leaps"; Parca makes things matter with her picks, as leaps, testing through transitions and shifts in "what might yet be/have been/could still have been [as tangible] experiments with their very being" (Barad 2012: 208). This is responsible theorizing on Parca's part if we think, like Barad (2012: 265), of responsibility "not [as an] obligation that the subject chooses . . . not as a calculation to be performed" but as "an incarnate relation that precedes the intentionality of consciousness . . . a relation always already integral to the world's ongoing intra-active becoming and not-becoming . . . the enabling of responsiveness."

Parca feels responsibility in the act of picking, in her pick intuitions. Everyone wants Parca to pick numbers for them. But Parca says to do so is *so much responsibility that it hurts*. If they lost a lot of money she would be devastated by the burden and feel like she would owe the losers for their losses, owe them a reason, owe them another chance, owe them her secrets. She can't afford that any way you cut it. That's one of the reasons why she is so careful not to make her picks public. She realizes now that that could be dangerous *all the way round*, she says. But there is more to the responsibility than this relationship to the money. It is the more intimate relations of blood, gristle, bone, and the forces of her ancestors, generations of dreamers, of wind and current, coral and sand, the weather, iguana antics, fish, family feuds, enemy talk, her work, the wild fragrance of her frangipani blossoms, the sweetness of the mangoes that grow in the yard—all of these connections need to be tended to with care and so these relations will care for her, through her numbers.

I once thought, half in jest, that Parca and I could create an "app" using her pick system and sell it through the BTL, "Belize Telecommunication Limited," the big phone company owned by Lord Michael Ashcroft. After all Gilroy "Press" Cadogen, the "sports psychic," writes a weekly Wednesday column on the Boledo and on sports, called "Press Predictions," for the *Amandala*, the national newspaper of record in Belize, and he's made quite a name for himself predicting Boledo numbers along with sports scores. Parca found nothing funny or useful in the suggestion. It left her with such an *angry stomach. And besides*, Parca said *Press doesn't say anything that anyone wouldn't already know anyway*, if they just attended to things.

Such an intimacy that sharing numbers creates can easily morph into a form of desperate accountability if not a precarious power over others or a convulsive power through them. It's happened before and it brought nothing but trouble, Parca remembers. Parca has much more than this history to deal with. *If I stick to my picks, that's more than enough*, she says. For Parca the responsibility of the pick is in what is yet to come in terms of an aesthetic of potential responsiveness. The power of such picking is not that kind of *power over* people and things. That would be too much like *a control over things*, she says and that is not anything like an act that promotes an enhanced sensibility to the enmeshed temporalities forming the ties of life in Wallaceville, in her social and material relationships that a pick generates. Responsibility here is a sustaining with-ness that Parca feels in a moment, a feeling with, a feeling how, feeling how far, how close, how else, as textures

Sunday, April 3, 2016   AMANDALABelize

PRESS PREDICTIONS
The sports psychic speaks

Press predictions — 40, 23, 57
Boledo stats thru 1st to 8th April, 2016
In boledo news, some sweet numbers will be out this week. These are my lucky numbers for this week – 40, 23, 57, 48, 11, 02. These are my lucky jackpots – 2848, 7234, 6510. These are my Pick-3 numbers – 014, 568, 973, 281, 603. Lots of doubles are coming out of that cylinder. The other 3 doubles that are overdue are 77, 11, 99. These numbers that are playing were long overdue. 03 is still out since the 5th of October, 2014, that's over 1 year and 6 months comes 5th of April. 62 is still out since the 18th of June 2014, that's over 1 year and 8 months now. 50 just joined those 2 numbers out over 1 year and more. 40 and 25 are around the curve. Get ready Belize; some dinero will be coming our way. My 85 played Sunday. One of those numbers will be out again. Remember to read your Psalms: 62, 92, 76. Good luck to all buyers. Next week more lucky numbers.

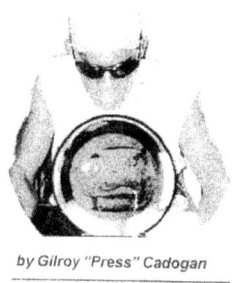

by Gilroy "Press" Cadogan

Figure 7.6. Gilroy "Press" Cadogan's Boledo picks for the week. ("Press Predictions," *Amandala*, 3 April 2016.)

that are inherent to the conjecture of experience and some emergent sense of a coming community, as Agamben (1993) puts it, and not only the human community, or in Rancière's (2004) words, some redistribution of the sensible that overthrows the regime of the perceptible as some swirling and unruly activity, some otherwise. For Parca it is always a new swerve, a fresh swirl, each a different figure of turbulent forces that is conjured in a pick, the pieces of which "are the various assemblages and individuals, each of which groups together an infinity of particles entering into an infinity of more or less interconnected relations" (Deleuze and Guattari 1986a: 254).

    Parca's picks always make her dizzy, it's a dizzy, giddy feeling she gets as she feels things shift, as new collaborations evoke new leaps, new stories bring her into touch, new forces of the sensible throw new number combinations and twists into relation: the swirling. Parca's picks are felt numbers, they are her becoming in touch with the world's aliveness, adding to it her energy. Parca's picks animate things and encourage a liveliness as an indeterminacy-in-action (see Barad 2012: 210). Her feely dream numbers are thrown together or falling apart; my try to move with Parca's picks through refrains of an intimate expressivity are a coming to pass through comings to be—a world in the act of its becoming by twisting expression out of matter taking shape in sinew and sense: it's a social in the making, an open-ended social (Massumi 2002a: 9). And in the act of Parca's picks a sociality emerges, composed of corporeal and incorporeal capacities, things thrown together—affects, hauntings, dreams, crazy things that happen during the day, failures, sturdy practices—as moments of endurance or sustainability, or not, nevertheless, moments understood as some twist and tangling swirl of elements and energies, morphing into something that may or may not actualize but seems always to open onto something else. Parca's picks are always of some assortment of the same stuff but each pick is a little bit different, something is added as a quality of becoming sustainable through all the little moments of touch, her life, then, a threshold of sustainability composed as a multiplicity of tiny uprisings (Berardi 2012).

Desire, Pick, Write

Parca's Boledo picks, my writing: how does it matter to write connected to Parca's stiff attempts to maintain traction in a confusing neoliberal present on the precipitous edges of empire, in a place called Belize, in the spaces that take place in the course of historical forces of eroding edifices, images, tendencies, and attachments to life? How does it matter to write tethered to Parca's picks as an emergent yet unregistered otherwise that is not about an object or a subject, but, rather, is itself a twist of repetition and difference composed of both potentialities and loss? Writing not as an expression of what is already acknowledged or known, but to write as an attending and an attuning, as a composing expression (Stewart 2007, 2008). How does it matter that I picked thirty-three and Parca did too? When that happened, some jumpy trajectory, some "relationscape" (Manning 2009), started to grow vital, fugitive, rogue, exuberant as some disjunctive self-inclusion (Massumi 2002a: 17). That's whoopy too, and I try to imagine what it might be like to lose myself in the currents of desire that are the very stuff of whoopy, in the unsettling passions of Parca's picks, in the endless uprisings of desire, in the delirium of an affective intensity conjured through numbers.

Picking numbers and writing: both practices gesture toward something generative in everyday moments of hunkering down, of dipping and dodging, of panic, rapture, deep curiosity, seduction or some eerie apotheosis that takes Parca's picking and my writing beyond some fraying "good life" tropical Paradise narrative of unsustainable "sustainable tourism" that some call Belize's "last resort," and toward some figure of an enduring late liberal otherwise (after Povinelli 2012). Here, impasse and potentiality rub up against each other as some double enactment, no ideological or material substratum, no "being behind doing effecting becoming" as Nietzsche (1967: 45) put it. Rather, it's the relational heterogeneity of Parca's picks as a transitory state of tension and seduction, as an intensity, an endurance in the way Bergson ([1910] 1950: 90–106) might put it, something that is durative, and for him that means something in the act of its potentializing, that draws my attending. The pick is everything to Parca, like the writing is to me.

This relational heterogenesis, call it *desire*, "for it is a matter of flows, of stocks, of breaks in the fluctuations of flows," Deleuze and Guattari (1986a: 105) write, something that "undoes its subjects, that exceeds its objects" (Pandian 2017), something that only wants to make things and break things. *After all*, Parca says, *I stay close to those feelings. Rough-up or make nice, I care for them*. She feels the material sensations that generate her numbers and cares for what is conjured in the pick through such intensities of sensation. Moving from one register and another, Parca's energies distribute across a field of subjects-objects-bodies-sensations-trajectories in the act of picking, in its eventfulness. These are assemblings that come into view as an ecology of activating passions through which things may be sent *swirling*, as Parca puts it. "Assemblages are passional, they are compositions of desire. . . . there is no desire but assembling, assembled desire. The rationality, the efficiency, of an assemblage does not exist without the passions the assemblage brings into play . . ." (Deleuze and Guattari 1986a: 399).

To attend to the activating passions of picking numbers is to stir up the various flows of Parca's picks as performances, dramas, and fieldings. This means lingering in the eventfulness of a number's milieu, numbers not just as the integers of an equation governed by big mathematical models of prediction and a calculated functionalist economic logic of Belizean financialization that would then render Parca's pick to be a local enactment of global, neoliberal frictions or a critical story on how Creole villagers understand money, precarity, and reproduction today. Parca's picks may be all of that or just plain lucky guesses, but they are something otherwise as well, the something that happens in the attraction or detraction of number sets, the something else that vibrates in the infinity of an emergent quality of movement in-between numbers, at thresholds (Massumi 2002a: 6–8). The otherwise of Parca's picks are held in those thresholds of life, incommensurate moving derangements of endurance, the swirlings that do not answer to Parca's name alone. The otherwise and the incommensurate are too transitory for that.

Neither is it the point to analyze Parca's picks as representations of some transforming world, some metaphor for life in some sprouting, out-of-control tourist Paradise. Rather it is to track picks as forces of expression and ask what generative modalities of knowledge, relationship, and attending linger in the picking as potentializing resonances that are extensive, entangling, and distributive, an emergence taking form in swirls, encouraged in the multiple relations between bodies, numbers, stuff, and dreams, where sensations and events are transduced through a vibratory relationality, a "response-ability" (Barad 2012: 208). I understand now that when Parca first demanded I *think about it* that this was another way of asking me to attend to how things compose themselves through number picks, how picking numbers sets things "swirling" into some contact infra-active (see Massumi 2015: 114) threshold of animate and inanimate, corporeal and incorporeal movement, movement taking form in the shifts that take place between the planes of dreams, sensations, and everyday events, a vibratory constellation that actualizes in Parca's number picks.

Of Boledo Picks and Poesis

Picking is Parca's way of exceeding the dismal dystopias of ordinary life lived in Wallaceville, this unsettled little place on the uncertain edges of imperial, global "digital-financial machines" now tied firmly to international tourism (Berardi 2012: 26). Picking contains the possibility of endurance for her. It is her act of reworking a life history of pain, disappointment, and fallen apart modes of cultural production while living bad-mouthed and beaten-up, and transforming this into a threshold of sustainability, an opening through things, fractured but holding, feeling something livable in the "right now" that doesn't equally feel like the world completely giving way again (but c.f. Nouvet 2014). For Parca feeling otherwise is an intuited sense of becoming present, an active becoming in picks, a creative not quite human assemblage as forces capable of producing something otherwise through the strange processes of vortices, eddies, and spirals of wild Caribbean currents: Parca's swirls.

Parca's picks and my writing are swirling enactments too: expressions energizing things. Such acts of expression can actualize the momentum of emergence, its uncaptured transitioning, just as this momentum recedes, diffuses, surges, and then repeats. Both are expressions in formation, the violence of their capture in code and representation at hand, yet right now, in the act of a pick, they are virtual, and remain unbound and singular, at the edges of that which has not yet been thought-felt. The trick is to catch expression as it is forming, as an act of expressivity, which is actually still an activating force of deformation. For Parca, picking Boledo numbers is a wrenching, stretching, and twisting "more than" of experience; picking activates certain contours of an event, it's becoming, it's movement, a complex matrix of sensations and sensibilities: the swirl. Here I work with the compositionality or the composability of Parca's picks trying in a manner of writing to stay attuned to the qualities and capacities of curiosity and care with respect to the potentialities of things happening through her picks to make up her life, to the dense and rogue expressivity of a life lived intimately in numbers, otherwise, or not.

The otherwise of Parca's picks enacts the conditions of possibility for new social life micro "uprisings" (Berardi 2012), forming as a "coming community" (Agamben 1993), a potentiality that depends on vitalities of force, human and nonhuman agencies, which together incite oscillating vibrations of unfolding compositions: fieldings. Parca's picks are her engagement with things as they are, her re-workings of endurance that help keep a body from cracking up, leaving open the question about whether there is always a way through things (Povinelli 2011). As such, Parca's picks incite a threshold of sustainability that is a fantasy of becoming as some capacity to imagine things incommensurate with the here and now. They are a series of connected affective intensities expressing a guess through a wide and wild sensory relay of energetic momentum: intuition. Here, the seductions of numbers and Parca's flirtation (see Crouch 2010; Ness 2011) with them are caught up in push and pull moments of excess and affect that ingathers things as a complex ecology of activating passions, practices that bring dreams, feelings, stuff and everyday events together in an affectively charged, multisensory relationship immanent to sensing corporealities in touch, as matter energizing the stuff of Parca's flashes.

But there is always more there (Thrift 2004: 59) that helps me avoid writing Parca's life in some "language of [personal and political] obituary" (Raffles 1999: 350). The more has to do with how numbers finally move things and bodies. The more is enlivened by the way Parca scans sensations absorbed in those off-shore fantasies and the local everyday chaos of a village that has gone crazy for tourism, in the way she reaches through her entire sensorium as a manner of inhabiting her body: follow the reef water, its temperature, its ever shifting blues, bad weather, good fishing, flotsam, the mangroves, the flows of sargassum that suddenly appear on a stiff breeze, the texture of blossoms on her old flamboyant tree, her sister's horrid words, the transient sweet tastes of mangoes, the antics of iguanas, bad accidents, the reverberate provocations of Obeah seductions, good luck, gecko chirps, tourist tricks, loves, labors, and daily dramas, the heat of the sand, all incite transitional passages: eventfulness as number emergence; desire in its transitory state.

I write in the gravitational orbit of these elementary particles (Houellebecq 2000) that add to things for Parca even if things never quite add up, a matrix of practices enacting a "doing" by which Parca and I begin to conjure our creation stories, me trying in a manner of writing to stay attuned to the qualities and capacities of curiosity and care with respect to the potentialities of whoopy, to the dense and wild generation of a life lived here, in this place, in the swirling restlessness of touchy numbers, on a tourist beach: a cultivation of desire in connectability. Parca enters into an erotics of the pick, in sensations of affliction and delectation working desire like a machine winding up but breaking, in all of those elements of conjuncture and disjuncture, unleashing impulses, flesh fluxes passing through Parca undoing her solidity. Desire, delirium, Parca conjuring her number picks passing from one intensity to another, from one event to another, from one dream and another, from one thing or another, the desiring machine undone by waves of delirium, suspensions, becomings, her leaps across voids. Parca's desire, much more than a personal feeling to fulfill as an object, is an incessant assembling that acts like waves do on the Wallaceville beach that flow and follow in relays, blockages, broken patterns, reforming figures and things always the same but different, confounding, surfacing, capping, folding, and those things that are unspeakable in waves of thick air.

"Desire," writes Deleuze, "never needs interpreting, it is it which experiments." Parca's picks are experiments in the play of forces, desire as poetic time-space assemblings that are the stuff of her life in Wallaceville today. Desire is a sort of serendipitous assembling, experience opening up as multiplicity, "contingent modes as a figure of fortunate errans" (Derrida 1980: 25). As such, desire is epiphantic and monstrous excess that, as expressed in Parca's picks, escapes the order of exchangeability and calculation for singular incitements of life. Such singularities are not compliant with any order of things but are her conditions of becoming in Wallaceville. Whoopy. This is how it goes day after day, wave of intensity after wave of intensity.

Parca's picks are co-composing intuitions, fieldings fashioned when a hard life rubs up against some sense of unspeakable possibility. Picking is a compositional practice that matters, the way writing matters, the way my storytelling reminds me continuously of the materiality of writing practices as knowledge-making practices, something out of nothing, the embeddedness of my writing within material processes that constitute and exceed them. It's the way Parca leans into things and how a world is thrown together or how it falls apart. That means my writing must stay lively in an effort to keep up with the agencies of what's being thrown together or falling apart. Either way Parca's picking practices are a fielding. They conjure a poetics of making a difference marked by how our senses might attune to storytelling in other modalities or stories told differently and in registers not eclipsed by binary thinking. Writing that makes a difference and so avoids referential and propositional logics that strive to keep things in their place by practices of disavowing and debunking. Picking numbers and writing are both practices of tracking where things might go if left unchecked by such logics. Here I think about extending the purview of expressivity beyond human-centered articulations and to envision it in terms of a generative multiplicity that resists binary modes of representation that try to name an already established world as its order of activity. Prefiguring a pick as a mathematical pre-existence, as a statistical formula and logic implies non-duration. It stops things in their tracks. But Parca endures and therefore, in that odd durative moment, she may be free. Bergson says that prefiguring is having an idea of a future act to be realized by effort, real but still undefinable. This is duration, a qualitative multiplicity, a heterogeneity (Bergson [1910] 1950: 204–26).

Parca's picks are difference generators through which she finds, or doesn't find, new paths to articulate relationships that register as new kinds of differences in the world and by so doing she makes a difference. For Parca difference is conjured in the spaces that are thick with the intimacies of her dreams, affective intensities and rogue events that swarm with energies, potentializing propositions, new formulations, expressed in a process and a pick. Writing too should be a difference generator. With Parca's invitation and inspired by her curiosity I have tried to draw different attention to the feel of co-composing through the creative intimacies of responsibility in a Caribbean tourism dreamworld-nightmare, on a beach, in Belize.

Epilogue
BELIZE FABULATIONS

What is planned in . . . fabulation are the conditions for an action as yet unmappable.
—Erin Manning, *The Minor Gesture*

The goal . . . is not to . . . develop an ontology of potentiality but to understand the dwelling of potentiality, asking what to do when we reach the limit of critical theory.
—Elizabeth Povinelli, "The Will to Be Otherwise/The Effort of Endurance"

Make-Belize Difference

Tourism is about change and movement. The kind of anthropology of tourism that I am writing, and the ethnographic methodology that sustains it, seems an ideal site to begin the task of creating an ethnography of difference, fashioned out of a reconsideration of change and movement through seven acts of Belize fabulation. With Belize fabulations *somet'ing* happens, and that *somet'ing in-forms* the make-Belize world of Wallaceville as an emergent world, one in the making, generative. With that, I tell my make-Belize stories while asking how such fabulations might augment the theory and methods usually employed in the anthropology of tourism, in tourism studies, and in anthropology more generally. Through Belize fabulations in the seven stories I tell, I invite readers to consider how a different form of writing, in this case make-Belize writing, may help open new, if uncertain, relations and channels for critical thought and creative expression that may then help contribute to a renewed sense of change and movement in the world, in *life*, in Belize. Make-Belize fabulations are the open-ended conjuring of an otherwise, the alternative possibilities of collective existence, of life taking form in Wallaceville. Wallaceville stories, lively and intense enough to reshape reality or to make something of *somet'ings*, are the "worlding" of a place gone *crazy, crazy* for tourism.

When I ask where encounters begin in Wallaceville I address the question from the middle of things where things are already in a process of change. So I begin not with the tourist, not with the local, neither with the "you" nor the "I," the meeting of the host and the guest, but in the infinite time-space of the in-between, where the forces of make-Belize activate things beyond everything decided. When things are decided it then only remains the job of recounting what has happened. The time-space of the in-between is that of transition, metamorphosis, incalculability, of aberrant monsters and grotesqueries, the noise of everything, castings and departures, swirling sea-sky emergence, indeterminacy, and generation, before things settle into the paths and structures that name them.

These seven stories of make-Belize linger in the chaos of the in-between that are meager attempts to instantiate a "world to come" in the throes of its becoming *somet'ing* that feels like *somet'ing*. They generate sensations and intensities as their mode of addressing problems. Ethnography as fabulation finds its life not in assessing concepts but in incitements and provocations of affecting and being affected and in the speculative resonances that affective intensities generate that urge life into new unfinished and generative worlds of relating, and so imagine human-nonhuman relations that are not beholden to the limited imperial political horizons of the Anthropocene and its connections to five hundred years of Caribbean colonial violence, extractive capitalism, and racism that anchors the hubris of a Euro-Imperial version of human exceptionalism and ecological collapse, more generally described by Haraway (2016) as the Capitalocene. Such storytelling as make-Belize fabulation also points to the epistemological challenges and contradictions contained in understanding the "imperial Caribbean" and a representationalist focus of Caribbean tourism studies, and tourism studies more generally, using conventional representationalist strategies of ethnographic realism, critically or scientifically applied (see McLean 2017a: 10–13; Myers 2018).

That is why I turn to "speculative fabulation" (after Haraway 2016: 12) as a way of inaugurating the forces of the otherwise, so to write in the cross currents of anthropology's imperial roots, evident in the representationalist strategies that still underwrite the most critical and collaborative of its efforts, so to establish the possibilities of historical narratives and political imaginaries of difference (see Said 1989: 209–10). While imperial formations of the Capitalocene imposed forms of life on colonized Belizeans that were and remain historically violent and racist, it is crucial to create ways to imagine connections across these forms so as to struggle with their conditions of possibility and the limitations of critical anthropology to address them.[1]

As I have said throughout the chapters of this book, these make-Belize stories are meant to add to things even if things never actually add up (Ivy 1995: 20; Stewart 2008: 72). That means that these stories act, and they are situations too, made-up activations from the "middle of things" that are much more than simple critical reflections on a given world of nature, politics, economy, and culture as collective representations. They are transformations too that are meant to be acts of "staying with the trouble" (Haraway 2016). And so, as something activating, these stories are much more than a collection of data about events, things real or make-believe, critical reflections, and complicated theory talk. These stories are meant to change the world with their telling. If ever so modestly, and I mean modestly, these inventions add to reality (Massumi 2002a: 13), hopefully like a poem does, no real use, maybe, but lots of value. I think of my stories akin to what McLean (2009) calls a "poetics of making (*poesis* in its most inclusive sense)." Stories are acts of "coming-to-being of the material universe," he says. As such, they are acts of creation, or what the very early modern magician-philosopher Giordano Bruno, in his creation of a "theatre of the world," called acts of *magic*, a complex dynamism, a uniquely open-ended ontogenesis. Deleuze calls this magic a "superior empiricism." Massumi calls it an "incorporeal materialism" (2002a: 5). I call it writing make-Belize stories.

Make-Belize stories are acts of encounter in their own right, and as such just like all acts of "becoming-with" encountering, they can make things happen (Haraway 2016: 12). Encounters and stories of encounter are both worlds in the making and as such they are part of life processes themselves, understood here as the entire field of relations as a world-forming process. "Becoming-with, not becoming, is the name of the game; becoming-with is how partners are . . . rendered capable. Ontologically heterogeneous partners become who and what they are in relational material-semiotic worldings. Natures, cultures, subjects and objects do not pre-exist their intertwined worldings" (Haraway 2016: 12–13).

What sustains my move to writing make-Belize tourist ethnography as a mode of anthropological thought and expression is not that it serves as an imaginative frill, some aesthetic embellishment, a hobbyist's conceit, an artistic diversion from the real facts and the serious business of transmitting authoritative knowledge about the real world, or navel-gazing reflexivity useful for a critical defense of make-Belize encounter stories as strong descriptions of a tropical "world on the wane." Ethically and politically the hope of this speculative ethnography is that it turns on initiatives that touch on new forces of responsibility that may help generate new forms of life demanding a new perspective on living and that this may inspire new acts of learning things differently, new relationalities, new experience beyond those forms of colonial and postcolonial agency that helped financialize everyday life, and new modes of attending to things beyond representational practices: possible make-Belize maneuverabilities (see Barad 2012; Haraway 2016; Massumi 2003).[2]

Monsters

While on a visit to Stockholm in late summer 2018, I spent a day in the Moderna Museet—Stockholm (The Modern Art Museum of Stockholm). It is Sweden's leading museum of modern and contemporary art. Moderna Museet collects and exhibits modern and contemporary art from around the world. The collection consists of art from the early twentieth century and photography from 1840. It was in the gallery of early twentieth-century art, right next to the extraordinary collection of political posters of the early Russian revolution, that I discovered *Monster* by the Cuban artist Fernandez Lourdes. The card associated with the painting read:

Fernandez Lourdes
f/b 1940
Kuba/Cuba
Monster, 1960
Akvarell och gouache
Inkop 1964

It is nearly impossible to describe the moment I first encountered *Monster*. With no definite article to anchor it, no "the" to hold it in place, Lourdes's *Monster* ebbed and flowed between mythical substance, nightmares, hallucinations, earth ooze and sea matter, and an infinity of forces and attributes through which, for me, it actualized as an unfolding make-Belize beast-time *somet'ing*. Caught in the force of its grip, *Monster* immediately transported me to Wallaceville and the Caribbean sea-sky creatures that Miss Grace sometimes talked about, morphing beasts and changelings that she said fishers would describe seeing in the waters inside the barrier reef just off the south coast of Belize. Not fish. Not manatee. Not turtle. Not dolphin. Not not-fish, manatee, turtle, or dolphin. Not human. Not not-human. Mermaid? Mer-monster? Not sea grass. Not not-sea grass. Not just sea creature but, yes, sea creature. *Somet'ing real*, Miss Grace said, *but not*.

Miss Grace once told me the story of the time in the 1990s when her Daddy said he saw one while fishing off Lark Caye. She never forgot his description of the *beast-ting* or the event of its appearance in her Daddy's story. Miss Grace's description of the *beast-ting* was remarkable. Unlike the named creatures of the world, this *beast-ting* was something real-unreal and its manifestation cast doubt on the taken-for-granted world of the Belize coast just at the time when all sorts of uncharacteristic things were beginning to take place. It was like encountering something that drifted, unanchored and dislodged of a particular species identity, unclassifiable, unknown, unrecognizable, but not.

The monsters that Miss Grace encountered, first in her Daddy's vivid story and then again when I showed her the Lourdes image, that she said left her *breathless*, remain unnamed but deeply felt. *Monster* and Miss Grace's *beast-ting* are aberrant interstitial and generative forms of activity in which all manner of life-forms are in transition, expressing a middle-of-things indeterminacy that she senses as beast-time matter forming, the felt and striking forces of an emergent make-Belize. They powerfully convey tidal waves of connecting forces, inchoate, indifferent, unspecified, and unstable proliferations that touch on the contingencies of connection and relating, of things assembling in Wallaceville: close entanglements of myths, animals, water, sand, sun, hard rains, bad currents, garbage, iguanas, accidents, secrets, Obeah curses, gales, idle talk, back talk, morning light, weather reports, soft sandy footsteps, wet rice, a history of break ups, pick-ups, and change, the end of the road, the Flood, the tourists, the wake of a boat, parrots, geckos, their chirps, Maya prayers, ruins, mangroves. Life in transition, composing: emergence, matter, chaos, relation, becoming, nature-culture, feeling, matter, matter feeling, potential, movement, change. The stuff of this set of swirlings variously gather like wave patterns meeting shifting currents, never the same twice but always in repetition, to create a plan for an image not yet mappable. And fabulations activate things yet to be mobilized and so remain a force of deviation, generative, and potentializing.

Monsters carry with them traumatic Paradise futures exposing other beast-time traumas of floods, hurricanes, ecological collapse, and wave upon waves of tourist and tourism projects of which Wallaceville is composed today. The monsters work on ecologies and bodies as a kind of unsettling emergency, one not solely connected to the everyday stuff of social life in Paradise or of sustainable biological life itself but to the very constitution of experience and expression as the making of life on a beach in Belize. Stay with the monsters, the *beast-tings*, Grace says, with those forces and powers that move things otherwise. With Grace I try and stay within the orbit and current of her stories and think of her acts of storytelling as special skills of make-Belize.

Lourdes's *Monster* and Grace's *beast'ting* precede individuation as a determinant and named organism that establish an identity. Grace and the *beast-ting*, together with Lourdes's *Monster*, generate an issuing forth not of an animal fully constituted but of the *somet'ings* that link the currents of a Caribbean "geophilosophy" of becoming, "knotting" beings and things—animal, vegetable, mineral—elemental creatures and strivings, the human and inhuman, the organic and inorganic, an assemblage composed of virtualities, events, and singularities as some cluster of impersonal forces "becoming life," as Deleuze (2001) puts it.

Life

Life in the making in a make-Belize Wallaceville, in seven stories: following Grace's example, I adopt Deleuze's (2005) concept of *a life*, what Manning (2016: 8) activates as the invention of new modes of "life-living." "Life-living," Manning says, "is a way of thinking life with and beyond the human, thinking life as more-than-human." Life here is a flux of lively contingent material-semiotic entanglements surging through unlimited existence. The encounter of "make-Belize" *somet'ings* evokes "life-living" as it *in-forms* an ecology of the insensible, or a sense as matter forming, an act of indeterminate relationality that asks (along with Haraway 2016; McLean 2017a; Myers 2018; Povinelli 2016; Tucker 2017; and Yusoff 2013: 208) "at every juncture *what else* life could be" (my emphasis, Manning 2016: 8). It's about how an insensible ecology as a life-orientation in the making, always in motion, is existence in transition. It's where *somet'ing* might lead if left unchecked, as a "minor gesture," a gesture toward a Deleuzian multiplicity of becoming-human that always exceeds containment and capture (Manning 2016: 8; Stewart 2003b). Think of monstrous *somet'ings* as "an inhuman life of matter forming the impersonal substrate of experience and subjectivity without being directly accessible to either, or indeed, to language, demanding that words be used as 'bait . . . fishing for whatever is not word'" (McLean 2018: 218).

My own troubling acts of make-Belize are generated out of my ethnographic research and writing about things happening in Wallaceville, things caught in circuits of action and reaction, that have impacts, that accrue in uneven swirlings of entangled associations, modes of attunement, attention, and attachment that generate life in Belize. In such moments of make-Belize, tourism remains uncaptured by defined concepts. Such moments remain immanent, in the realm of the insensible. Make-Belize moments fold and unfold as potentializing story build-up, they are actively generative, in a state of emergence that makes them present, shifty, rogue, eccentric, distrait, hostile, uncompromising, but always unfinished in their heaving efforts to create the time and spaces of their realization. It is to the contingent, open-ended movement of things present in the acts of their emergence as "changing changes" and as "emergence emerges" that moves this work along (Massumi 2002a: 10). Processes of make-Belize, out of "curiosity, risk, potentiality and exhaustion," open onto a space of endurance, an "otherwise," that Povinelli addresses when she asks "what to do when we reach the limit of critical theory" (2012: 454). The work that these seven stories do, and the way that they do it, have been my initial attempt to address that question.

What also inspires the direction of this work may be best summed up by Donna Haraway when she implores us to "stay with the trouble" and this may be a provisional kind of sympathetic reply to Povinelli's question. "In fact, staying with the trouble," she says, "requires learning to be truly present, not as a vanishing pivot between awful or Edenic pasts and apocalyptic or salvific futures, but as mortal critters entwined in myriad unfinished configurations of places, times, matters and meanings" (Haraway 2016: 1). The uneasy, edgy parallels and rogue discontinuities between make-Belize and the make-believe inspire my stories. Through them troubling tangles of discontinuous forms of living throw themselves together as *somet'ing* make-Belize as the viscous, proliferative mess of a transitive what is, right now. It is in my concern with staying present and with the trouble of the unfinished, entangled, forever-changing processes of make-Belize, processes that entwine me in its poesis—the fabulation—that this book matters.

Through the processes of make-Belize invention, things modulate, affect, and transform each other. The materializing metamorphosis of acts of Wallaceville make-Belize are the processes that transform one half-formed historical thing into networks of other half-formed things as if by some crazy conjuring trick right before everyone's eyes: an apocalypse out of chaos, the narrowing specification of chaos from a particular point of disorder and unpredictability, a profusion of forces gathering and intensifying, a materiality mustering a measure and a rhythm out of nature without norm, into *somet'ing*, until the next *somet'ing* begins to happen. Attending to these ebb and flow processes as newly forming rhythms of life, a temporality with its newly textured and unsettling sensations and new forces and energies, I encounter the beast-time in Wallaceville.

My hope is that these make-Belize stories of *life* becoming *somet'ing* in the Belize beast-time can help with this task, help me ever so modestly pull anthropological thinking about Belize and tourism out of its own stubborn desire for some unspoiled equivalence between an analytical subject, a confidence activated with the concepts used to take on the subject, and the world—"a kind of razed earth for academic conversation," as Stewart (2008: 72) puts it, that establishes tourism analysis in terms of the discourse of critical values or techno-fixes (Haraway 2016; but see Ness 2016; Tucker 2017; Ren 2010).

As such, this writing is meant to incite an ecology of the insensible as one that activates passions as a kind of attunement to the emergent potential of make-Belize. Such an ecology is instantiated through language when we think of language as "something" written in the same way we think of a bit of beach sand, a blade of sea grass, the buzz of an insect, the rustling of coconut trees in the breeze in which the earth and, what Deleuze calls, the deeply creative nature of the material world continues to "matter" (see Grosz 2008). Mattering is "sense as matter forming, as cohabitation," a creativity beyond the creativity that is habitually attributed to society, culture, and mind (Yusoff 2013: 208). It is the difference that language as matter itself makes, as opposed to the difference that consciousness makes of matter, that makes my writing a generative practice and a palpable presence. The materiality of language as a medium, words in the shape of sound, breath, vibration, cadence are the matter of worldings that writing strives to engage, working with powers and intensities present in language that are always at work in the world. Writing, as an act of make-Belize with its emphasis on the matter and materiality of expression, means activating ethnography's powers as a participatory activity of "becoming-with" worldings.

Fabulate

Life in the beast-time make-Belize begins with a once upon a time "Flood" story. Ecological disaster, the act of hurricane winds and downpour rain and a roiling quake that opened the earth and swallowed Wallaceville whole. Flood, winds, and quake connecting a disturbing rise in sea levels (melting ice caps) and water temperatures (rising CO_2 levels), dangerously shifting currents, disappearing beaches, the barrier reef bleaching out, fish gone, cayes going, cruise ship garbage of mostly plastics accumulating and smothering beach and sea, while shiny oils smother all life in greasy slicks and ruin, a viscous mess: a mad proliferation of trouble.

This is another "once upon a time," and there are many more connected to it, so that a field of stories without firm limits but made of lots of different possibilities *in-form*s a worlding and makes an opening for thought that activates the possible rather than describes the probable. The best story to tell is the one that enlivens the processes of entanglement, "not the one that already carries within itself its own fix" (Manning 2016: 8). Stories of the make-Belize beast-time summoned through stories of that alarming Flood that somehow summoned those waves of immigrant populations of retirement-age Europeans and North Americans with big claims on a small and vulnerable village and country. Putting on the pressure, too many tourists with dangerous demand for at-home comforts and service with a smile, a demand for local and immigrant laborers to build their beach palaces, cook their meals, provide their security and the menial labor for upscale holiday and retirement comfort. *Paradise*, the tourists call it. *Apocalypse*, Miss Grace snaps back, her bible firmly in hand, as she makes an Obeah curse that summons more possible connections and conjunctions: the pro-generative power of one beastly Flood to conjure another monstrous beast, the transformative quality of Xtabay's inhuman species unanchored, greedy seedy seductions, instantiating a beast-time flood of tourists and a sun bleached, bloated, and outrageous dreamworld phantasmagoria to become impossible Paradise promises.

There is legitimizing power in remaking Belize as a neoliberal "tourist state," and then studying it as a fully named subject, the governmentality of which operates aesthetically through the management of the representational machinery of a tropicalized Paradise, one way or the other. Belize tourism can be said to be an aesthetic of containment through romantic tropical images, bodies, nature, and culture, all commodified forms of phantasmagorical tourism aesthetics that transforms both material and bodily dimensions of Wallaceville, numbing and desensitizing both tourists and the toured through exposure and the cleaver management of excessive sensory stimulation. It keeps everyone in the loop and anchored to an image. In Belize today the phantasmagoric aesthetic at work is operationalized through tourist scenes (tours, images, performances, etc.) and sites (resorts, nature stagings, enframed ruins) and the way in which such scenes and sites reproduce themes of captured nature, culture, and bodies that circulate as immediate realities of a well governed and disciplined tourist state.

My intervention into this controlled containment site is to fabulate (Flaxman 2012). I repeat: fabulation is an act of creation that emerges in the middle of things that are meant to connect and to add to things coming and going as "another way of traveling and moving," becoming (Deleuze and Guattari 1986a: 25). "[C]oming and going rather than starting and finishing," my seven stories move between things like the movements of Caribbean currents and waves that seem to have no beginning or end but endlessly repeat themselves as they pick up speed and intensity or dissipate and disappear, as they re-contour the shoreline (Deleuze and Guattari 1986a: 25). My stories sweep one way then another, up and down, and as such they are meant to resist the temptation common to tourism studies to contain things at a stand-still for critical inspection. Fabulations are emergent processes of becoming and they act like the endless forces of wave action, the same but different, always in motion refashioning things. As such, fabulations are lively entanglements of human and nonhuman forces that deal in becomings, if becoming involves being carried over and beyond the very possibility of a contained identity, a subject and an object.

Miss Grace traveled past her father floating on churning waves in his coffin out to sea while she floated in the opposite direction, holding to her dislodged house, toward the lagoon. Her face contorted. Anguish and shock. Chaos confusion. She shake violence. Covered in wet and bruise. Eyes dilated. Hair in greasy coils. Muscles cramped for holding on. Parched throat. Bile. Guts churning. Fire fever hot in cold Flood plague. She scream her prayer with a defiant Obeah curse! Big gulps of filth and air. *Hold'in baibl.* Delirious. *Hold'in on.* And the awful sound of her Daddy's grizzle laugh. No man-sound like it. Only trickster monster, boney death-life. Dark and violent force, now and forever more. *Beast-time 'pon wi all*, Miss Grace recalls.

Miss Grace trembles with the recall. The recall repeats and spreads uncontained like a virus with every *crazy crazy* tourist *t'ing*, a *beast-ting*, that happens now and makes life impossible. And every time she recalls the *beast-ting* and Xtabay—those half-human, half-animal figures of unfixed gender, transmuting metamorphosis, and elemental juxtapositions from mixed times and places—and what Xtabay did with her Daddy, she knows that her *time is at hand*. *Like mi baibl say*, she spits. Her time out of Flood time, now the beast-time middle of things, things created churning out of so many other things carried on current, wave, and wake. She feels it. Dizzy dissolution. Undifferentiation transforming into the ugly hardened specter of a tourist Paradise. *Good for noth'in*, Miss Grace say. Richie agrees, Parca too. *Wi watch di beast feed on life here now*, Parca spurts. Xtabay seduces with those new pots of tourist lucre, and then consumes life whole making things and people *crazy crazy*. Like the tourists, like the locals, like the sea, like the sand, like the wind, like the Flood. What else? Green snake seductions. What else? All a story, not true, not false, not good, not evil. Nothing ever is. Instead, seven stories. Speculative fabulation, experimental all the way down, here through seven stories that make a difference through acts of invention. Like sea waves, each story makes a cut, each cut makes a difference; *what else* becoming *somet'ing*.

These seven stories are affirmations, each a minor gesture, each not true but not false, by which I mean these stories find creative potential in remixing, becoming, and challenging the given order of things as the *potencia* (creative power) of metamorphosis (Braidotti 2002: 21). These stories are "projection machines." As such they address "a people yet to come" and so they are not formed in individual or collective identity but in the otherwise of tourist, long- and short-term visitor, expat, and local encounters in Wallaceville, their desires and becomings (Deleuze and Guattari 1999: 230). The storytelling of "a people yet to come" stresses the "what else" of my seven stories fashioned through historical narrative fragments that together generate the seductive and demanding forces of experience in Wallaceville today. These are the fragments out of which the make-Belize is always in the act of composing. *Somet'ing* becoming what else.

This book, therefore, through the stories it tells, engages the anthropology of tourism with a literature that it mostly doesn't think with, namely critical posthumanist and new materialist work in ontogenesis, concerned with how encounters "matter" and how matter is thought and constituted through vibrant entanglements, refrains, knots, and figures of human and nonhuman bodies, affects, objects, and practices. Attending to tourism studies thusly reorients thinking around questions of relationality, about the resonances of material-semiotic forces co-implicated in what bodies can "do" and how "matter" "acts" rather than a concern with what "is" a body or the agentic meaning of experience when considering the relational processes of encounter. Here I am interested in how tourism activates potentialities in bodies to be otherwise, to generate certain kinds of Paradise natures, mutations, and affects as insensible natures, as the agitation or provocation and curiosity and desire that draws over the work of intelligibility in acts of encounter. So the insensible draws its energy as an agitation in movement what Grosz calls "nicks in time" or Barad thinks of as "quantum cuts" or Deleuze and Guattari call "virtual ecologies" or Massumi calls a "rhythm without regularity." These are focused indeterminacies, in the example of Caribbean Belize sea-sky swirlings, and attention to them might possibly energize thinking through encounters in tourism as indeterminate processes that remain otherwise and incommensurate with given forms of knowledge production that need not any longer anchor the study of encounters in tourism studies. My hope is that more tourist and travel ethnography might begin to take seriously the arts of fabulation in the age of the Capitalocene and so learn to conjure something livable and not just describe the living. Such art I believe would encourage us to hone skills to help fashion a politics of care while cultivating livable worlds, opening up the life powers of emergent . . . "what else."

Notes

1. Here, Amitav Ghosh's novels, especially his *Ibis* trilogy, including *Sea of Poppies* (2008), *River of Smoke* (2011), and *Flood of Fire* (2015); the work of Michael Jackson, especially his *Excursions* (2007); and the novels of Camilla Gibb, especially her *Sweetness in the Belly* (2005): all trained field anthropologists, who, each in their own way, imaginatively invent transoceanic movements of displaced lives across time, space, and desire (see Stankiewicz's 2012 interview with Ghosh).
2. Lots of "new," I know. Ethnographic writing has always been a site of political and poetic struggle, intimacy, experiment, and change (see Clifford and Marcus 1986). But what is different about this work is that it envisages writing as a force of passage that equips us for a move to think otherwise, to bend curiosity so that it becomes untethered from given practices of critical attention and thought "to follow the objects it encounters, or become undone by its own attention to things that don't just *add up* but take on a life of their own as a problem of

thought" (Stewart 2008: 72). See Biehl and Locke 2017; McLean 2017; Myers 2018; Pandian and McLean 2017a; Stewart 1996, 2008; and Taussig 1997, 1999, 2003, 2006, 2009, 2011.

There is a fast-growing interest in experimental writing in anthropology and allied arts. I have highlighted some of the best work in anthropology that is related in one way or another, critically or affirmatively, with the work of Deleuze and Guattari and appeal to those in the anthropology of tourism especially, to learn how to enjoy writing differently with these experimenters. Why? Because it is with them that writing becomes a way of conceiving life yet to come (Flaxman 2012). Life yet to come asserts the minor gestures of speaking with not speaking about or for, as a matter of "making kin," conjuring the forces of entanglements and a shared relationality, however tenuous or contingent, and tracking those forces, their creative powers, as an act of living through other people, places, things unalike and letting them live through the stories we tell and, so, move beyond. This is a "becoming with" that is also a "becoming sensor in sentient worlds" (Myers 2017: 73). This is also what Povinelli (2012) has called "the will to be otherwise," the tenuous and fraught effort to track knowledge as a project of ethical transformation, an experiment in attuning to the world yet to come, in its becoming, in its unfinished transformations.

There are other anthropologists whose creative writing efforts follow other philosophical trajectories and practices not necessarily connected to Deleuze, yet have no less powerfully led to a new awareness of the importance of fiction, performance, and experience. I point especially to those who consider themselves to be storytellers in the rich traditions of auto-ethnography, intertextuality, and reflexive-critical writing of all sorts that necessarily are meant to engage with and transform collective life, creating a generativity of things as they are emergent rather than remain apart from them in description and analysis. The rich work of Ruth Behar (1996, 2007, 2013), Paul Stoller (2008, 2016), and Helena Wulff (see her 2016 edited volume on the anthropologist as writer) are exemplary. The sources referenced are my favorites. They reach into several rich ethnographic traditions, most recently critical reflexive anthropology and phenomenology. Some write through a critical hermeneutic tradition of inventive interpretation from which Clifford Geertz found deep inspiration in adopting "fiction" as his method, or the life history methodology that harkens back to Paul Radin's early critical ethnography of the Winnebago (See Little 1981) or the generative forms of storytelling that are the formative practices for creating particular cultural lives, especially through the memoir or ethnopoetics, to name a few. A turn to these traditions has disrupted a privileging of anthropological knowledge production and ethnographic authority in order to encourage anthropologists, lured by fiction, to open their writing to imaginative worlds more attuned to dialogue, polyphony, resonance, narrative, and experience, and so develop a critical politics and poetics of ethnographic encounters.

Glossary

I am only representing translations of Kriol words as they were used by my interlocutors in my conversations with them.

All translations and transcriptions are based on the *Kriol-Inglish Dikshineri/English-Kriol Dictionary* (Herrera, Manzanares, Woods, Crosbie, and Decker 2007). In general nouns are uninflected for number and possession and verbs are uninflected for tense, person, and number (see Decker 2005). Note that Belize Creole designates the ethnic group and Belize Kriol designates the language spoken by this group.

Please also note that Kriol is the most widely used of the many languages spoken in Belize, even though the lingua franca of the country is English. I do not speak Kriol and have only a rudimentary understanding of its grammar and vocabulary. Thus, under many circumstances, my Creole interlocutors would code switch to English when they felt that I did not understand the rich stories they were telling me. This is most apparent in the stories I recorded for Chapter 1. I have done what I can to render these stories intelligible while at the same time capturing the socio-performative-imaginative forces that underwrite their production and telling.

a: of
aai ya yai: form of exclamation
ada: other
afta: after
an: on
baibl: bible
bak: back
bering grong: graveyard
buk: book

bwai: boy
chrash: trash
dehdeh: there
den: they
devl: devil
doti: dirty
dotinis: filth
fi: for
fi chroo: truly
fi dehn: their
gaad: god
gat: have
goh: go
gwehn: going to
gyaabij: garbage
gyal: girl
hapi: happy
jompi: jumpy
kaafin: coffin
kalek: collect
kip: keep
kloas: close
kolcha: culture
kom: come
lampa: lamp post, with reference to a lazy person, indolence, hanging around
lat: a lot
mangro: mangrove
mek: make
naitmyaa: nightmare
nuh: no
oava: over
prais: price
prayaa: prayer
rass: harassment, bullshit
renk: stink
sens: sense
stammi: stormy
ting: thing
tink: think
vais: voice
vexed: disturbed

Glossary

wach: watch out
wan: want
weh: way
yuh: you

References

Agamben, Giorgio. 1993. *The Coming Community*, trans. Michael Hardt. Minneapolis: University of Minnesota Press.

———. 1998. *Homo Sacer: Sovereign Power and Bare Life*, trans. Daniel Heller-Roazen. Stanford: Stanford University Press.

Ahmed, Sara. 2010. *The Promise of Happiness*. Durham, NC: Duke University Press.

Anderson, Ben. 2009. "Affective atmospheres." *Emotion, Space and Society* 2: 77–81.

Bærenholdt, Jorgen Ole, Michael Haldrup, Jonas Larsen, and John Urry, eds. 2004. *Performing Tourist Places*. Aldershot: Ashgate.

Bakhtin, Mikhail. 1981. "Forms of Time and of the Chronotope in the Novel." In *The Dialogical Imagination. Four Essays*, ed. Carly Emerson and Michael Holquist, 84–258. Austin: University of Texas Press.

Bakke, Gretchen, and Marina Peterson, eds. 2017. *Between Matter and Method: Encounters in Anthropology and Art*. London: Bloomsbury.

Barad, Karen. 2003. "Posthumanist Performativity: Toward an Understanding of How Matter Comes to Matter." *Signs: Journal of Women in Culture and Society* 28(1): 801–31.

———. 2012. "On Touching—the Inhuman That Therefore I am." *Differences: A Journal of Feminist Cultural Studies* 25(3): 206–23.

Behar, Ruth. 1996. *The Vulnerable Observer: Anthropology that Breaks Your Heart*. New York: Beacon Press.

———. 2007. *An Island Called Home: Returning to Jewish Cuba*. New Brunswick: Rutgers University Press.

———. 2013. *Traveling Heavy: A Memoir in Between Journeys*. Durham, NC: Duke University Press.

Benitez-Rojo, Antonio. 1990. *Sea of Lentils*. Amherst: University of Massachusetts Press.

———. 1992. *The Repeating Island: The Caribbean and the Postmodern Perspective*, trans. James Maraniss. Durham, NC: Duke University Press.

Benjamin, Walter. 1968. *Illuminations, Essays and Reflections*, trans. H. Zohn. New York: Schocken Books.

———. 1999. *The Arcades Project*, trans. H. Eiland and K. McLaughlin. Cambridge, MA: The Belknap of Harvard University Press.

Bennett, Jane. 2010. *Vibrant Matter: A Political Ecology of Things*. Durham, NC: Duke University Press.

Berardi, Franco "Bifo." 2012. *The Uprising: On Poetry and Finance*. Cambridge, MA: Semiotext(e)/MIT Press.

Bergson, Henri. (1910) 1950. *Time and Free Will: An Essay of the Immediate Data of Consciousness*, trans. F.L. Pogson. London: George Allen and Unwin Ltd.

———. (1912) 1991. *Matter and Memory*, trans. Nancy Margaret Paul and W. Scott Palmer. New York: Zone Books.

———. 2007. *The Creative Mind: An Introduction to Metaphysics*, trans. Mabelle L. Andison. Mineola, NY: Dover Books.

Berlant, Lauren. 2007. "Slow Death (sovereignty, obesity, lateral agency)." *Critical Inquiry* 33(Summer): 754–80.

———. 2011. *Cruel Optimism*. Durham, NC: Duke University Press.

Berlant, Lauren, and Michael Warner. 1998. "Sex in Public." *Critical Inquiry* 24(2): 547–66.

Biehl, João. 2005. *Vita: Life in a Zone of Social Abandonment*. Berkeley: University of California Press.

Biehl, João, and Peter Locke. 2017. *Unfinished: The Anthropology of Becoming*. Durham, NC: Duke University Press.

Blanchot, Maurice. 1997. *Friendship*. Stanford: University of Stanford Press.

Braidotti, Rosi. 2002. *Metamorphoses: Towards a Materialist Theory of Becoming*. Cambridge, UK: Polity Press.

Brennan, Denise. 2004. *What's Love Got to Do with It: Transnational Desires and Sex Tourism in the Dominican Republic*. Durham, NC: Duke University Press.

Bruner, Edward. 2004. *Culture on Tour: Ethnographies of Travel*. Chicago: University of Chicago Press.

Buck-Morss, Susan. 1991. *The Dialectics of Seeing: Walter Benjamin and the Arcades Project*. Cambridge, MA: MIT Press.

———. 1992. "Aesthetics and Anaesthetics: Walter Benjamin's Artwork Essay Reconsidered." *October* 62: 3–41.

———. 2000. *Dreamworld and Catastrophe: The Passing of Mass Utopia in East and West*. Cambridge, MA: MIT Press.

Buda, Dorina Maria. 2015. *Affective Tourism: Dark Routes in Conflict*. London: Routledge.

Castaneda, Quetzil. 1996. *In the Museum of Maya Culture: Touring Chichén Itzá*. Minneapolis: University of Minnesota Press.

Christ, Sally, Seleni Matus, and Vincent Palacio, eds. 2001. *National Tour Guide Training Program: Trainers Manual*. Belize City: Genesis Arts.

Clifford, James, and George Marcus, eds. 1986. *Writing Culture: The Poetics and Politics of Ethnography*. Berkeley: University of California Press.

Colman, Steven, and Michael Crang. 2002. "Grounded Tourists, Traveling Theory." In *Tourism: Between Place and Performance*, ed. S. Colman and M. Crang, 1–20. Oxford: Berghahn Books.

Crick, Malcolm. 1994. *Resplendent Sites, Discordant Voices: Sri Lanka and International Tourism*. London: Routledge.
Crouch, David. 2010. *Flirting with Space: Journeys and Creativity*. Aldershot: Ashgate.
Cvetkovich, Anne. 2003. *An Archive of Feeling: Trauma, Sexuality, and Lesbian Public Cultures*. Durham, NC: Duke University Press.
Dave, Naisargi. 2011. "Indian and Lesbian and What Came Next: Affect, Commensuration, and Queer Emergences." *American Ethnologist* 38(4): 650–65.
Debord, Guy. 1995. *The Society of the Spectacle*. New York: Zone Books.
Decker, Ken. 2005. *The Song of the Kriol: A Grammar of the Kriol Language of Belize*. Belize City, Belize: Belize Kriol Project.
Deleuze, Gilles. 1997. *Essays Critical and Clinical*. Minneapolis: University of Minnesota Press.
———. 2001. *Pure immanence: Essays on a life*, trans. Anne Boyman. New York: Zone Books.
———. 2005. *Francis Bacon: The Logic of Sensation*, trans. D.W. Smith. Minneapolis: University of Minnesota Press.
Deleuze, Giles, and Félix Guattari. 1986a. *A Thousand Plateaus: Capitalism and Schizophrenia*, trans. Brian Massumi. Minneapolis: University of Minneapolis Press.
———. 1986b. *Kafka: Toward a Minor Literature*, trans. Dana Polan. Minneapolis: University of Minnesota Press.
———. 1999. *What is Philosophy?*, trans. H. Tomlinson and G. Burchell. London: Verso Press.
Derrida, Jacques. 1980. *Writing and Difference*, trans. Alan Bass. Chicago: University of Chicago Press.
Destination Belize: The Official Visitor Magazine of the Belize Tourism Industry Association. 2003. Belize City: The Belize Tourism Board.
Dhillon, Bob, and Fred Langan. 2018. *Business and Retirement Guide to Belize: The Last Virgin Paradise*, 2nd ed. Toronto: Dundurn Press.
Di Giovine, Michael A. 2014. "The Imaginaire Dialectic and the Refashioning of Pietrelcina." In *Tourism Imaginaries: Anthropological Approaches*, ed. Noel B. Salazar and Nelson H. H. Graburn, 147–171. Oxford: Berghahn Books.
Duffy, Rosalind. 2000. "Shadow Players: Ecotourism, Development, Corruption, and State Politics in Belize." *Third World Quarterly* 21(3): 549–65.
Elliott, Denielle, and Dara Culhane, eds. 2017. *A Different Kind of Ethnography: Imaginative Practices and Creative Methodologies*. Toronto: University of Toronto Press.
Feldman, Joseph. 2011. "Producing and Consuming 'Unspoiled' Tobago: Paradise Discourse and Cultural Tourism in the Caribbean." *Journal of Latin American and Caribbean Anthropology* 16(1): 41–66.
Flaxman, Gregory. 2012. *Deleuze and the Fabulation of Philosophy*. Vol. 1 of *Powers of the False*. Minneapolis: University of Minnesota Press.
Franklin, Adrian, and Mike Crang. 2001. "The Trouble with Tourism and Tourism Travel Theory." *Tourist Studies* 1(1): 5–22.
Frohlick, Susan. 2013. "Intimate Tourism Markets: Money, Gender, and the Complexity of Exotic Exchange in a Costa Rican Caribbean Town." *Anthropological Quarterly* 86(1): 133–162.

———. 2016. "Feeling Sexual Transgression: Subjectivity, Bodily Experience, and Non-Normative Hetero-Erotic Practices in Women's Cross-Border Sex in Costa Rica." *Gender, Place & Culture* 26(2): 257–73.

Frohlick, Susan, and Julia Harrison. 2008. "Engaging Ethnography in Tourist Research: An Introduction." Special issue, *Tourist Studies* 8(5): 5–18.

Fullagar, Simone. 2001. "Encountering Otherness: Embodied Affect in Alphonso Lingis' Travel Writing." *Tourist Studies* 1(2): 171–83.

Ghosh, Amitav. 2008. *Sea of Poppies*. Toronto: Viking Canada.

———. 2011. *River of Smoke*. Toronto: Viking Canada.

———. 2015. *Flood of Fire*. Toronto: Viking Canada.

Gibb, Camilla. 2005. *Sweetness in the Belly*. Toronto: Doubleday Canada.

Graburn, Nelson. 1983. "The Anthropology of Tourism." *Annals of Tourism Research* 10(1): 9–33.

———. 2001. "Secular Ritual: A General Theory of Tourism." In *Hosts and Guests Revisited: Tourism Issues of the 21st Century*, ed. V.L. Smith and M. Brent, 43–50. New York: Cognizant Communication.

Gray, Bill, and Claire Gray. 1999. *Belize Retirement Guide: How to Live in the Tropical Paradise on $450.00 a Month*, 4th ed. Toronto: Preview Publishing.

Gregg, Melissa, and Gregory J. Seigworth. 2010. "An Inventory of Shimmers." In *The Affect Theory Reader*, ed. Melissa Gregg and Gregory J. Seigworth, 1–25. Durham, NC: Duke University Press.

Grosz, Elizabeth. 2008. *Chaos, Territory, Art: Deleuze and the Framing of the Earth*. New York: Columbia University Press.

Hagerty, Timothy, and Mary Gomez Parham, eds. 2000. *If Di Pin Neva Ben: Folktales and Legends of Belize*. Mexico City: Cubola Productions.

Hall, Norris. 1974. *Grinning Doukies: A Look at Lottery*. Personal papers. Belize National Archives.

Haraway, Donna. 2008. *When Species Meet*. Minneapolis: University of Minnesota Press.

———. 2016. *Staying with the Trouble: Making Kin in the Chthulucene*. Durham, NC: Duke University Press.

Hardt, Michael, and Antonio Negri. 2000. *Empire*. Cambridge, MA: Harvard University Press.

Harrison, Julia. 2003. *Being a Tourist: Finding Meaning in Pleasure*. Vancouver: University of British Columbia Press.

Herrera, Yvette, Myrna Manzanares, Silvana Woods, Cynthia Crosbie, and Ken Decker, eds. 2007. *Kriol-Inglish Dikshineri/English-Kriol Dictionary*. Belmopan, Belize: Belize Kriol Project.

Holmes, Teresa. 2010. "Tourism and the Making of Ethnic Citizenship in Belize." In *Tourism, Power and Culture*, ed. Donald V.L. Macleod and James G. Carrier, 153–73. Bristol: Channel View Publications.

Houellebecq, Michel. 2000. *The Elementary Particles*. New York: Alfred A. Knopf.

Hutnyk, John. 1996. *The Rumour of Calcutta: Tourism, Charity, and the Poverty of Representation*. London: ZED Books.

Ivy, Marilyn. 1995. *Discourses of the Vanishing: Modernity, Phantasm, Japan*. Chicago: University of Chicago Press.
Jackson, Michael. 2007. *Excursions*. Durham, NC: Duke University Press.
Johnson, Melissa A. 2019. *Becoming Creole: Nature and Race in Belize*. New Brunswick, NJ: Rutgers University Press.
Kaplan, Caren. 1996. *Questions of Travel: Postmodern Discourses of Displacement*. Durham, NC: Duke University Press.
Kempadoo, Kamala. 2004. *Sexing the Caribbean: Gender, Race and Sexual Labor*. New York: Routledge.
Kirshenblatt-Gimblett, Barbara. 1998. *Destination Culture: Tourism, Museums, and Heritage*. Berkeley: University of California Press.
Kohn, Eduardo. 2013. *How Forests Think: Toward an Anthropology Beyond the Human*. Berkeley: University of California Press.
Lacan, Jacques. 1977. *Ecrits: A Selection*. New York: W.W. Norton.
Leite, Naomi. 2014. "Afterward. Locating Imaginaries in the Anthropology of Tourism." In *Tourism Imaginaries: Anthropological Approaches*, ed. Noel B. Salazar and Nelson H.H. Graburn, 260–78. Oxford: Berghahn Books.
Little, Kenneth. 1981 "Explanation and Individual Lives: A Reconsideration of Life Writing in Anthropology." *Dialectical Anthropology* 5: 215–26.
———. 1991. "On Safari: The Visual Politics of a Tourist Representation." In *The Varieties of Sensory Experience: A Sourcebook in the Anthropology of the Senses*, ed. David Howes, 149–63. Toronto: University of Toronto Press.
———. 2010. "Paradise from the Other side of Nowhere: Troubling a Troubled Scene of Tourist Encounter in Belize." *Journal of Tourism and Cultural Change* 8(1,2): 1–14.
———. 2014. "Mr. Richie and the Tourists." *Emotion Space and Society* 12: 92–100.
———. 2012. "Belize Blues." *Semiotic Inquiry* 32(1,2,3): 25–46.
MacCannell, Dean. 1976. *The Tourist: A New Theory of the Leisure Class*. New York: Schocken Books.
———. 1992. *Empty Meeting Ground: The Tourist Papers*. London: Routledge.
Manning, Erin. 2007. *Politics of Touch: Sense, Movement, Sovereignty*. Minneapolis: University of Minnesota Press.
———. 2009. *Relationscapes: Movement, Art, Philosophy*. Cambridge, MA: MIT Press.
———. 2011. "Fiery, Luminous, Scary." *SubStance* 40(3): 41–48.
———. 2016. *The Minor Gesture*. Durham, NC: Duke University Press.
Massumi, Brian. 1992. *A User's Guide to Capitalism and Schizophrenia: Deviations from Deleuze and Guattari*. Cambridge, MA: MIT Press.
———. 1993. "Everywhere You Want to Be: Introduction to Fear." In *The Politics of Everyday Fear*, ed. Brian Massumi, 3–38. Minneapolis: University of Minnesota Press.
———. 2002a. *Parables for the Virtual: Movement, Affect, Sensation*. Durham, NC: Duke University Press.
———. 2002b. *A Shock of Thought*. New York: Routledge.

———. 2003. "Navigating Moments: An Interview with Brian Massumi." In *Hope: New Philosophies for Change*, ed. M. Zournazi, 210–43. New York: Routledge.
McLean, Stuart. 2009. "Stories and Cosmologies: Imagining Creativity Beyond 'Nature' and 'Culture.'" *Cultural Anthropology* 24(2): 213–45.
———. 2011. "Black Goo: Forceful Encounters with Matter in Europe's Muddy Margins." *Cultural Anthropology* 26(4): 589–619.
———. 2017a. *Fictionalizing Anthropology: Encounters and Fabulations at the Edges of the Human*. Durham, NC: Duke University Press.
———. 2017b. "Sea." In *Crumpled Paper Boat: Experiments in Ethnographic Writing*, ed. Anand Pandian and Stuart McLean, 148–67. Durham, NC: Duke University Press.
———. 2018. "It." In *Posthuman Glossary*, ed. Rosi Braidotti and Maria Hlavajova, 216–21. London: Bloomsbury Academic.
Medina, Laurie Kroshus. 2003. "Commoditizing Culture: Tourism and Maya Identity." *Annals of Tourism Research* 30(2): 353–68.
Meloy, Ellen. 2003. *The Anthropology of Turquoise: Reflections on Desert, Sea, Stone, and Sky*. New York: Vintage Books.
Mitchell, Timothy. 1988. *Colonising Egypt*. Cambridge, UK: University of Cambridge Press.
Myers, Natasha. 2017. "Becoming Sensor in Sentient Worlds: A More-than-natural History of a Black Oak Savannah." In *Between Matter and Method: Encounters in Anthropology and Art*, ed. Gretchen Bakke and Marina Peterson, 73–96. London: Bloomsbury Academic.
———. 2018. "How to Grow Livable Worlds: Ten Not-So-Easy Steps." In *The World to Come: Art in the Age of the Anthropocene*, ed. Kerry Oliver-Smith. Gainesville: Samuel P. Harn Museum, University of Florida.
Nail, Thomas. 2018. *Lucretius I: An Ontology of Motion*. Edinburgh: Edinburgh University Press.
Nelson, Maggie. 2009. *Bluets*. Seattle: Wave Books.
Ness, Sally. 2011. "Flirting with Boulders: Changing Configurations of Landscape in Yosemite National Park." *Journal of the Association of Social Anthropologists Online* 1(6): 1–59.
———. 2016. *Choreographies of Landscape: Signs of Performance in Yosemite National Park*. Oxford: Berghahn Books.
Nietzsche, Fredrich. 1967. *The Genealogy of Morals*, trans. Walter Kaufmann and R. J. Hollingdale. New York: Vintage Books.
Nouvet, Elysee. 2014. "Some Carry On, Some Stay in Bed: (In)Convenient Affects and Agency in Neoliberal Nicaragua." *Cultural Anthropology* 29(1): 80–102.
Ngai, Sianne. 2004. *Ugly Feelings*. Cambridge, MA: Harvard University Press.
Ochoa, Todd Ramon. 2007. "Versions of the Dead: Kalunga, Cuban-Kongo Materiality, and Ethnography." *Cultural Anthropology* 22(4): 473–500.
———. 2017. "Origami Conjecture for a Bembe." In *Crumpled Paper Boat: Experiments in Ethnographic Writing*, ed. Anand Pandian and Stuart McLean, 172–84. Durham, NC: Duke University Press.

Pandian, Anand. 2012. "The Time of Anthropology: Notes from the Field of Contemporary Experience." *Cultural Anthropology* 27(4): 547–71.

———. 2017. "Desire in Cinema." In *Crumpled Paper Boat: Experiments in Ethnographic Writing*, ed. Anand Pandian and Stuart McLean, 119–25. Durham, NC: Duke University Press.

Pandian, Anand, and Stuart McLean, eds. 2017a. *Crumpled Paper Boat: Experiments in Ethnographic Writing*. Durham, NC: Duke University Press.

———. 2017b. "Prologue." In *Crumpled Paper Boat: Experiments in Ethnographic Writing*, ed. Anand Pandian and Stuart McLean, 1–10. Durham, NC: Duke University Press.

Picard, David, and Michael A. Di Giovine, eds. 2014. *Tourism and the Power of Otherness: Seductions of Difference*. Bristol: Channel View Publications.

Picard, Michel. 1996. *Bali: Cultural Tourism and Touristic Culture*, trans. Diana Darling. Singapore: Archipelago Press.

Piot, Charles. 2010. *Nostalgia for the Future: West Africa after the Cold War*. Chicago: University of Chicago Press.

Pollard, Mariam. 1989. *The Listening God*. Wilmington, DE: M. Glazier.

Povinelli, Elizabeth. 2006. *Empire of Love: Toward a Theory of Intimacy, Geneology and Carnality*. Durham, NC: Duke University Press.

———. 2011. *Economies of Abandonment: Social Belonging and Endurance in Late Liberalism*. Durham, NC: Duke University Press.

———. 2012. "The Will to be Otherwise/The Effort of Endurance." *South Atlantic Quarterly* 111(3): 453–75.

———. 2014. "Geontologies of the Otherwise." Editors' Forum: Theorizing the Contemporary, 13 January. https://culanth.org/fieldsights/geontologies-of-the-otherwise.

———. 2016. *Geontologies: A Requiem to Late Liberalism*. Durham, NC: Duke University Press.

Pratt, Mary Louise. 1992. *Imperial eyes: Travel writing and transculturation*. London: Routledge.

Proust, Marcel. 2006. *Remembrance of Things Past*, vol. 1, trans. C. K. Scott Moncrieff. Hertfordshire, UK: Wordsworth Editions, Ltd.

Rabelais, Francois. 1955. *Gargantua and Pantagruel*, trans. Sr. T. Urquahart and P. Motteux. London: Encyclopedia Britannica, Inc.

Raffles, Hugh. 1999. "'Local Theory': Nature and the Making of an Amazonian Place." *Cultural Anthropology* 14(3): 323–350.

Rancière, Jacques. 2004. *The Politics of Aesthetics: The Distribution of the Sensible*, trans. Gabriel Rockhill. London: Continuum.

Ren, Carina. 2010. "Constructing Tourism Research: A Critical Inquiry." *Annals of Tourism Research* 37(4): 885–904.

Roessingh, Carel, and Karin Bras. 2003. "Garifuna Settlement Day: Tourism Attraction, National Celebration Day, or Manifestation of Ethnic Identity." *Tourism, Culture and Communication* 4: 163–72.

Rojek, Chris, and John Urry, eds. 1997. *Touring Cultures: Transformation of Travel and Theory*. London: Routledge.

Sade, Marquis de. 1966. *The 120 Days of Sodom*. New York: Grove Press.
Said, Edward W. 1989. "Representing the Colonized: Anthropology's Interlocutors." *Critical Inquiry* 15(2): 205–25.
Salazar, Noel. 2010. *Envisioning Eden: Mobilizing Imaginaries in Tourism and Beyond*. Oxford: Berghahn Books.
———. 2012. "Tourism Imaginaries: A Conceptual Approach." *Annals of Tourism Research* 39(2): 863–82.
———. 2013. "Imagineering Otherness: Anthropological Legacies in Tourism." *Anthropological Quarterly* 86(3): 669–96.
Salazar, Noel B., and Nelson H. H. Graburn, eds. 2014. *Tourism Imaginaries: Anthropological Approaches*. Oxford: Berghahn Books.
Saldanha, Arun. 2007. *Psychedelic White: Goa, Trance, and the Viscosity of Race*. Minneapolis: University of Minnesota Press.
Scott, David. 1999. *Refashioning Futures: Criticism after Postcoloniality*. Princeton: Princeton University Press.
Scott, Julie, and Tom Selwyn, eds. 2010. *Thinking through Tourism*. Oxford: Berg Publishers.
Seatone Consulting. 2015. "A Social Viability Assessment of Cruise Tourism in Southern Belize." San Francisco: Seatone Consulting. http://www.seatoneconsulting.com/wp-content/uploads/2015/02/BelizeCruiseViabilityAssessment2011.pdf.
Sharpe, Christina. 2016. *In the Wake: On Blackness and Being*. Durham, NC: Duke University Press.
Sheller, Mimi. 2004a. "Demobilizng and Remobilizing Caribbean Paradise." In *Tourism Mobilities: Places to Play, Places in Play*, ed. Mimi Sheller and John Urry, 13–21. London: Routledge.
———. 2004b. *Consuming the Caribbean: From Arawaks to Zombies*. London: Routledge.
Shepard, Robert. 2018. "Cosmopolitanism and Tourism in a Post-Hegelian Age." In *Cosmopolitanism and Tourism: Rethinking Theory and Practice*, ed. Robert J. Shepard, vii–xxvii. Lantham: Lexington Books.
Simoni, Valerio. 2016. *Tourism and Informal Encounters in Cuba*. Oxford: Berghahn Books.
Sluder, Lan. 2010. *Easy Belize: How to Live, Retire, Work and Buy Property in Belize, the English Speaking, Frost Free Paradise on the Caribbean Coast*. Asheville: Equator Publications.
Smith, Valene. 1989. *Hosts and Guests: The Anthropology of Tourism*, 2nd ed. Philadelphia: University of Pennsylvania Press.
Solnit, Rebecca. 2005. *A Field Guide to Getting Lost*. New York: Penguin Books.
Stankiewicz, Damian. 2012. "Anthropology and Fiction: An Interview with Amitav Ghosh." *Cultural Anthropology* 27(3): 435–541.
Stengers, Isabelle. 2005. "The Cosmopolitical Proposal." In *Making Things Public: Atmospheres of Democracy*, ed. B. Latour and P. Weibel, 994–1003. Cambridge, MA: MIT Press.
Stewart, Kathleen. 1996. *A Space on the Side of the Road: Cultural Poetics in an "Other" America*. Princeton: Princeton University Press.

———. 2003a. "Arresting Images." In *Aesthetic Subjects: Pleasures, Ideologies, and Ethics*, ed. P. Matthews and D. McWhirter, 443–48. Minneapolis: University of Minnesota Press.

———. 2003b. "A Perfectly Ordinary Life." In "Public Sentiments: Memory, Trauma, History, Action," ed. A. Cvetkovitch and A. Pellegrini. Special issue, *The Scholar and Feminist On-Line* 2(1), Summer. Retrieved 25 May 2009 from http://www.barnard.edu/sfonline/.

———. 2007. *Ordinary Affects*. Durham, NC: Duke University Press.

———. 2008. "Weak Theory in an Unfinished World." *Journal of Folklore Research* 45(1): 71–82.

———. 2010a. "Atmospheric Attunements." *Environment and Planning D: Society and Space* 29(3): 445–53.

———. 2010b. "Worlding Refrains." In *The Affect Theory Reader*, ed. M. Gregg and G.J. Seigworth, 339–54. London: Duke University Press.

———. 2017. "Epilogue." In *Crumpled Paper Boat: Experiments in Ethnographic Writing*, ed. Anand Pandian and Stewart McLean, 225–30. Durham, NC: Duke University Press.

Stoller, Paul. 2008. *The Power Between: An Anthropological Odyssey*. Chicago: University of Chicago Press.

———. 2016. *The Sorcerer's Burden: The Ethnographic Saga of a Global Family*. London: Palgrave Publishing.

Sutherland, Ann. 1998. *The Making of Belize: Globalization on the Margins*. Westport: Bergin and Garvey.

Swain, Margaret Byrne. 2014. "Myth Management in Tourism's Imaginariums: Tales from Southwest China and Beyond." In *Tourism Imaginaries: Anthropological Approaches*, ed. Noel B. Salazar and Nelson H. H. Graburn, 103–124. Oxford: Berghahn Books.

Taccone, Ines. 2019. "The Feeling of the Fall: A Marginal Belize Story." PhD dissertation draft. Toronto: York University.

Taussig, Michael. 1993. *Mimesis and Alterity*. New York: Routledge.

———. 1997. *The Magic of the State*. Chicago: University of Chicago Press.

———. 1999. *Defacement: Secrecy and the Labor of the Negative*. Stanford: Stanford University Press.

———. 2003. *My Cocaine Museum*. Chicago: University of Chicago Press.

———. 2006. *Walter Benjamin's Grave*. Chicago: University of Chicago Press.

———. 2009. *What Color is the Sacred?* Chicago: University of Chicago Press.

———. 2011. *I Swear I Saw This: Drawings in Fieldwork Notebooks, Namely My Own*. Chicago: University of Chicago Press.

Thomas, Lynnell. 2014. *Desire and Disaster in New Orleans: Tourism, Race, and Historical Memory*. Durham, NC: Duke University Press.

Thompson, Kathleen. 2006. *An Eye for the Tropics: Tourism, Photography, and the Framing of the Caribbean Picturesque*. Durham, NC: Duke University Press.

Thrift, Nigel. 2004. "Intensities of Feeling: Towards a Spatial Politics of Affect." *Geografiska Annaler. Series B. Human Geography* 86(1): 57–78.

Tsing, Anna Lowenhaupt. 2004. *Friction: An Ethnography of Global Connection*. Princeton, NJ: Princeton University Press.

———. 2015. *The Mushroom at the End of the World: On the Possibility of Life in Capitalist Ruins*. Princeton, NJ: Princeton University Press.

Tucker, Hazel. 2017. "Contaminated Tourism: On Pissed Off-ness, Passion and Hope." Invited Plenary Lecture, Critical Tourism Studies VII, Mallorca, Spain, June 2017.

Urry, John. 1990. *The Tourist Gaze: Leisure and Travel in Contemporary Societies*. London: Sage.

"What You Need to Know about Retiring in Belize." nd. *International Living*. Retrieved 25 September 2019 from https://signup.internationalliving.com/X120K C35/belize/belize-retirement.

Wilk, Richard. 1993. "'It's Destroying a Whole Generation': Television and Moral Discourse in Belize." *Visual Anthropology* 5: 229–44.

———. 1994. "Colonial Time and TV Time." *Visual Anthropology Review* 10(1): 94–105.

———. 1995. "Learning to be Local in Belize: Global Systems of Common Difference." In *Worlds Apart: Modernity through the Prism of the Local*, ed. Daniel Miller, 240–66. London: Routledge.

———. 2002. "Television, Time, and the National Imaginary in Belize." In *Media Worlds: Anthropology on New Terrain*, ed. F. Ginsberg, L. Abu-Lughod, and B. Larkin, 171–86. Berkeley: University of California Press.

———. 2006. *Home Cooking in the Global Village: Caribbean Food from Buccaneers Ecotourists*. Oxford: Berg Publishers.

Wulff, Helena, ed. 2016. *The Anthropologist as Writer: Genres and Contexts in the Twenty-First Century*. Oxford: Berghahn Books.

Yusoff, Kathryn. 2013. "Insensible Worlds: Postrelational Ethics, Indeterminacy, and the (K)nots of Relating." *Environment and Planning D: Society and Space* 31: 208–26.

INDEX

Note: page numbers in italics refer to illustrations; some people and events named in the index are fictional composites of people, events, and experiences.

affect, definition of, 20–21, 50, 78, 100–101
affirmation, definition of, 11, 171
Agamben, Giorgio, 77, 155, 159
Ahmed, Sara, 111
Aiden, 98
Alice, 94, 97
Anderson, Ben, 90
Anthropocene, the, 7, 163
anticolonial imagery, 9–10
applied tourism studies. *See under* tourism studies
Ashcroft, Michael, 8, 154
assemblage, definition of, 91, 157

Bakhtin, Mikhail, 92
Bakke, Gretchen, 22n3
Barad, Karen, 14, 146, 153–54, 155, 158, 172
Behar, Ruth, 173n2
Belikin Beer calendar girls, 66, 94
Belize, origin of name, 22n2
Belize Brewing Company, 86, *87*
Belize Tourism Board, 6–7, 8, 31, 75, 112
Belize Tourism Industry Association, 120
Belize Jewels, 94–95, 103
Benitez-Rojo, Antonio, 5, 25, 92
Benjamin, Walter, 93
Berardi, Franco "Bifo," 158, 159

Bergson, Henri, 16, 92, 127, 139, 157, 161
Berlant, Lauren, 33, 70, 88, 121, 139
Biehl, Joao, 6, 11, 22n3
Blanchot, Maurice, 14
Bob, 51–52
Bobby, 43–44
bodies, 14, 18–19, 29, 64, 65–66, 71, 77, 85, 110, 125–26, 172; local, 6, 63, 93, 109, 128, 131–32; tourist, 6, 14, 75–78, 82, 93–94. *See also* Belize Jewels; Doug; expats; Gaile; Half-Jack; Miss Grace; Parca; Royal Windsors, the
Boledo, 2, 127–28, 129–33, *130, 134–36,* 137, 146–49, 154, *155,* 158–61; history of, 128–30, 131
Book of Revelations, 25, 26, 27, 29, 30, 31, 52
Bosun Jack, 68
Bradley, Leo, 22n4
Braidotti, Rosi, 171
Bruno, Giordano, 19, 164
Buck-Morss, Susan, 77, 78, 79

Cadogen, Gilroy "Press," 154, *155*
Capitalocene, the, 6, 163
Captain Morgan's Resort and Casino, 108
Captain Ted, 101–2

Index

Cedar, 61. *See also* Happy Hour Hank, 61
Chas, 95, 96
chronotopes, 92, 129
colonialism, 6, 8, 22n2, 128–29, 142, 163
communication technologies, 7–8, 24n9, 33, 101
Conrad, Joseph, *The Children of the Sea*, v
consumer capitalism, 15, 20–21, 81, 85, 107, 119, 122; Belize as product for, 9, 10, 11, 77–78, 83, 95–97, 115
contact zone, concept of, 20, 63, 107. *See also* milieu
Coppola, Francis Ford, 8, 39
cosmopolitics, 87, 102
Cousteau, Jacques, 108
Crang, Mike, 12, 13
Crazy Mojo, 38–39
Creole, definition of, 22n4
critical tourism studies. *See under* tourism studies
cruise ship industry: expansion in Belize, 7, 9, 38, 54, 66, 99, 101, 112, 115–16, 122; garbage and environmental impacts of, 18, 83–84, 169
Culhane, Dara, 22n3
Cvetkovich, Ann, 97

Danny, 41–42
Debord, Guy, 77–78
Deleuze, Gilles, 28, 40, 71, 92, 146, 155, 164, 169, 170, 171; and affect, 50, 100; on assemblages, 157; concept of a life, 70, 167; on desire, 157, 160; on the field of immanence, 138, 152, 153; on milieux, 20n4, 63, 107; minor literature, 64, 65; and virtual ecologies, 14, 172; and writing, 1, 13, 16, 173n2
Derrida, Jacques, 160
Destination Belize (magazine), 47, 75, 108, 112
Di Giovine, Michael A., 90–91, 103
Diane, 51–52
Double-Jack, 68
Doug, 118–24

drugs, 4, 8, 29, 48, 76, 83, 84–85, 91, 93, 112, 114, 121; and anti-drug legislation, 79; and cartels, 99, 101, 131; local stories about, 45, 57, 66, 79–80, 92, 100, 123; and Richie's money, 43–44, 47, 61
Drunken Pirate, the, 98, 101

ecotourism, 6, 9, 17, 23n5, 28, 29, 54, 108, 113–14, 116, 120. *See also* international tourism industry; Paradise, creation of
Elliott, Denielle, 22n3
Elvis, 80
ephemera, 86–104; definition of, 92–93
ethnic tourism, 10
ethnographic writing, 16, 18–19, 21, 22n3, 162, 164–65, 172n2. *See also* speculative fabulation, definition of; tourism studies
evangelicals. *See* missionaries
expats, 17–18, 48, 52, 61, 65, 82, 84–85, 89, 90, 99, 131, 148, 153, 171; attacks on, 44–45, 66; come-go, 23n5; come-stay, 23n5; developments for, 10, 36, 80, 81, 119; long-stay, 23n5, 80. *See also* Alice; Bosun Jack; Captain Ted; Double-Jack; Doug; Gaile; Half-Jack; Hank; Miss Grace; Skip; Wally

fishing economy, decline of, 17, 38, 117, 169
Flood, the, 25
Franklin, Adrian, 12, 13

Gaile, 118–24
Ghosh, Amitav, 172n1
Gibb, Camilla, 172n1
Graburn, Nelson H.H., 90
Gregg, Melissa, 100
Grosz, Elizabeth, 14, 63, 107, 172
Guattari, Félix, 1, 28, 71, 92, 155, 157, 170, 171; on the field of immanence, 138, 152; and virtual ecologies, 14, 172; and writing in anthropology, 173n2

188

Index

Hagerty, Timothy, 27
Half-Jack, 67–72
Happy Hour Hank, 57–61
Haraway, Donna, 88, 106, 107; speculative fabulation, 4, 5, 14, 102, 163; on staying with the trouble, 1, 15, 164, 168
Harriet, 97
Heather, 80
Heidegger, Martin, 17
hippies, 75–76, 112
Hurricane Iris (The Flood), 2, 3, 25–31

international tourism industry, 33, 35, 39, 44, 54, 99, 102, 158; critiques of, 10; impacts of, 18, 62; marketing of Belize to, 6–9. *See also* cruise ship industry; ecotourism; land developers; Paradise, creation of
Ivy, Marilyn, 164

Jackson, Michael, 172n1
James, Marlon, v
Janet. *See also* Half-Jack, 68, 69, 71, 72, 73
Jimmy, 96, 97–98, 99–100
Jonestown, Guyana, 52
Joseph, 78
June, 43, 44

Kohn, Eduardo, 18

Lacan, Jacques, 102
Lamarche, Bernard, 53
land developers, 9, 33, 37, 38, 44, 51, 68, 80, 118, 131, 133. *See also* international tourism industry
Leite, Naomi, 89, 90, 102
Locke, Peter, 6, 11, 22n3
Lourdes, Fernandez, *Monster*, 165, 166
Love FM, 8, 39, 54, 129–30

Manning, Erin, 3, 4, 13, 67, 73, 146, 156, 169; on affirmation, 11; on fabulation, 162; on life-living, 167; on sensing beyond security, 15, 68, 88
Massumi, Brian, 16, 18-19, 71, 120, 156, 157, 158, 167; affirmative

augmentation, 85; on creative contagion, 50, 100–101; on developing analysis, 12; on examples, 49, 85; and fields of contact, 63, 67, 95, 107; and the force of expression, 65; on incorporeal materialism, 46, 47, 164; on intensity, 78, 88, 110; on intuition, 138, 139, 146; on rhythm without regularity, 14, 35, 172; and singularities, 98; on writing, 105
Matus, Melenie, 94
McAfee, John, 8
McLean, Stuart, 6, 22n3, 88, 163, 164, 167–68
Meloy, Ellen, 105, 117
migrant labor, 37, 66–67, 169
milieu, concept of, 20, 63, 106, 107. *See also* contact zone, concept of
Miss Gloria, 48
Miss Grace, 25–31, 32, 33, 36, 38, 39, 40–42, 165–66, 169–70, 170–71
Miss Jane, 80
missionaries, 17, 129, 133
Mitchell, Timothy, 17
Mr. G., 39, 140
Mr. Normal, 79
Mr. Pete, 44–45, 47–49, 50, 51
Myers, Natasha, 173n2
mystery ship, the, 98–102

National Tour Guide Training Program Manual, 22n4
neoliberalism, 7–9, 85, 131–32, 170
Ness, Sally, 6, 18
Ngai, Sianne, 17
NGOs, 8, 36, 85, 98, 131, 132, 133; ecotourist and environmental, 17, 61, 66, 116, 120, 131
Nietzsche, Fredrich, 157

Obeah spiritual practices, 36, 38, 42, 129, 149, 159; and curses, 26, 31, 166, 169–70
Ochoa, Todd Ramon, 4, 89
otherwise, the, 4, 14–15, 16, 18, 28–29, 40, 92, 146, 163, 167–68, 172n2.

189

See also Parca; speculative fabulation, definition of

Pandian, Anand, 6, 22n3, 35, 157
Paradise, creation of, 2, 4, 6–7, 29, 33–42, 44, 47, 170, 172. *See also* ecotourism; international tourism industry
Parca: and Boledo, 127–61, *130, 134–36*; and the fire at Ruby's restaurant, 139–43; and the Grinning Doukie Skeleton, 143, 144, *145*; on tourism and tourists, 171; and the visa interview, 147–50
Parham, Mary Gomez, 27
Pastor G, 152
Pastor Pat, 66–67
Peterson, Marina, 22n3
pirates: in Belize history, 75–76, 112, 128–29; stories about, 45, 91, 92, 108
poesis, 106, 119, 164, 168
Pollard, Miriam, 124
Povinelli, Elizabeth, 4, 18, 92, 162, 168, 173n2
Proust, Marcel, 86

Rabelais, Francois, 70
Raelians, 1, 8, 21n1, 47, 51–54, *53*, 55
Raffles, Hugh, 159
Rancière, Jacques, 155
resort entrepreneurs and development, 7, 29, 31–32, 33, 36, 37, 39, 44, 61, 115, 131. *See also* Bob; Captain Morgan's Resort and Casino; Coppola, Francis Ford; Diane
resort workers, 49. *See also* Parca
Richie, 1, 51; and drug money, 43–44, 47; and the fire at Ruby's restaurant, 139–40; on The Flood, 3; on Parca's Boledo picks, 137; on tourism and tourists, 22n2, 48, 52–53, 54–55, 57–74, 171; and the Tree of Truth, 54–55; on Twitch's killing, 79
Ricky, 49
Rita, 58, 60. *See also* Happy Hour Hank
Ron, 113–17
Royal Windsors, the, 76–77
Ruby, 139, 140, 142, 150, 153

Sade, Marquis de, 70
Said, Edward, 163
Salazar, Noel B., 90, 91, 102–3
Saldanha, Arun, 91
Seigworth, Gregory J., 100
Sergeant Ramos, 48
Skip, 147–49, 149–50, 152
Sluder, Lan, 9
Solnit, Rebecca, 106, 109
speculative fabulation, definition of, 11, 16–17, 102, 162, 163, 166
Stengers, Isabelle, 102
Stewart, Kathleen, 56, 62, 106, 122, 144, 164; and assemblages, 90, 91, 104, on emergent vitalities, 17, 91–92; on ethnographic writing, 16, 64, 89, 107, 168, 172n2; on singularities, 103; on worlding refrains, 109, 125
Stoller, Paul, 173n2
Stretch, 32, 114–15
Sutherland, Ann, 7, 8, 10, 79
Swain, Margaret Byrne, 90–91, 103
Sweet Sunsets Bar and Grill, 93–96
Sweets, 97, 115

Taussig, Michael, 86, 103
Thrift, Nigel, 15, 144, 159
timeshares, 9, 23n5, 39, 83
tourism studies, 90, 103, 162, 163, 170; applied, 12, 23n7; critical, 10–14, 22n3, 172
Toycie, 128
Tsing, Anna Lowenhaupt, 11, 15, 20
Twitch, death of, 1, 78–81, 84, 85

Vernon, 49

Wallace, Peter, 22n2
Wally, 129, 147
Warner, Michael, 70
World Bank, 8, 132
Wulff, Helena, 173n2

Xtabay, 27, 28, 30, 31, 32, 33, 36, 41, 42, 170, 171

Yusoff, Kathryn, 13, 167, 169

www.ingramcontent.com/pod-product-compliance
Lightning Source LLC
Chambersburg PA
CBHW070042040426
42333CB00041B/2110